APPLYING RADIATION SAFETY STANDARDS IN RADIOTHERAPY

SAFETY REPORTS SERIES No. 38

APPLYING RADIATION SAFETY STANDARDS IN RADIOTHERAPY

JOINTLY SPONSORED BY THE
EUROPEAN SOCIETY FOR
THERAPEUTIC RADIOLOGY AND ONCOLOGY
INTERNATIONAL ATOMIC ENERGY AGENCY,
INTERNATIONAL LABOUR OFFICE,
INTERNATIONAL ORGANIZATION FOR MEDICAL PHYSICS,
PAN AMERICAN HEALTH ORGANIZATION,
AND WORLD HEALTH ORGANIZATION

INTERNATIONAL ATOMIC ENERGY AGENCY
VIENNA, 2006

COPYRIGHT NOTICE

Sales and Promotion Unit, Publishing Section
International Atomic Energy Agency
Wagramer Strasse 5
P.O. Box 100
A-1400 Vienna
Austria
fax: +43 1 2600 29302
tel.: +43 1 2600 22417
http://www.iaea.org/books

Printed by the IAEA in Austria
February 2006
STI/PUB/1205

IAEA Library Cataloguing in Publication Data

Applying radiation safety standards in radiotherapy / jointly Sponsored by the International Atomic Energy Agency ... [et al.]. — Vienna : International Atomic Energy Agency, 2006.
p. ; 24 cm. — (Safety reports series, ISSN 1020-6450 ; no. 38)
STI/PUB/1205
ISBN 92–0–110904–0
Includes bibliographical references.

1. Radiation — Safety measures — Standards. 2. Radiotherapy — Safety measures — Standards. I. International Atomic Energy Agency. II. Series.

IAEAL 06–00417

FOREWORD

The International Basic Safety Standards for Protection against Ionizing Radiation and for the Safety of Radiation Sources (BSS) cover the application of ionizing radiation for all practices and interventions and are, therefore, basic and general in nature. Users of radiation sources have to apply these basic requirements to their own particular practices. This requires a degree of 'interpretation' by the user, which can result in varying levels of regulatory compliance and inconsistencies between applications of the BSS to similar practices. In this context, the preamble of the BSS states that: "The [regulatory body] may need to provide guidance on how certain regulatory requirements are to be fulfilled for various practices, for example in regulatory guideline documents."

In order to guide the user to achieve a good standard of protection and to achieve a consistent national approach to licensing and inspection, some countries have developed practice specific regulatory guidance, while others have practice specific regulations. For obvious reasons, national regulatory guidance is tailored to a country's own legislation and regulations. This can lead to problems if the guidance is used in other States without appropriate modification to take local requirements into account. There would therefore appear to be scope for producing internationally harmonized guidance, while bearing in mind that the ultimate responsibility for the regulatory documents rests with the State.

Some regions have taken the initiative of preparing guidance to facilitate the regional harmonization of regulatory control of certain common practices (e.g. radiotherapy). A number of draft regulatory guidance documents for the main practices involving the use of ionizing radiation have already been prepared. This initiative indicates that there is a global demand for such documents. In particular, it is felt that countries participating in the IAEA's technical cooperation model project on Upgrading Radiation and Waste Safety Infrastructure would benefit significantly from the availability of practice specific guidance. Member States could then more readily develop their own guidance tailored to their own requirements and needs. This idea led to the development of the present report.

The Action Plan on the Radiological Protection of Patients, approved by the IAEA General Conference in September 2002, requires that "The practice-specific documents under preparation should be finalized as guidance rather than regulations, and they should include input from professional bodies, from international organizations and from authorities with responsibility for radiation protection and medical care." Following this request, the only mandatory statements of this report are quotations from the BSS, including requirements.

There are certain BSS requirements that, when applied to specific practices, can be fulfilled mainly through one practical solution. In these cases, the regulatory body may need to use a 'should' statement, which implies that licensees should choose this solution or, if another option is intended, an equivalent level of safety should be provided. In other cases, there may be more than one option. In these cases the regulatory body would just mention or describe them.

This guidance is intended for both regulators and users of radiation sources in radiotherapy. Regulators may use it for reviewing applications for authorization and during the inspection of facilities. Registrants/licensees may wish to follow the guidance in order to comply with BSS requirements or equivalent national requirements. Experts recruited on IAEA missions to advise on the implementation of the BSS for the practice of radiotherapy are expected to use this regulatory guidance report rather than their own national guidance. Working safely, with a quality assurance programme, is important and contributes to gaining overall confidence and credibility in the practice of radiotherapy.

This report has been prepared by the IAEA with the contributions of the Pan American Health Organization, the International Labour Office, the World Health Organization, the European Society for Therapeutic Radiology and Oncology and the International Organization for Medical Physics.

The IAEA officer responsible for this publication was P. Ortiz López of the Division of Radiation, Transport and Waste Safety.

CONTENTS

1. INTRODUCTION

1.1. BACKGROUND

The International Basic Safety Standards for Protection against Ionizing Radiation and for the Safety of Radiation Sources (the 'Standards' or 'BSS') were published as IAEA Safety Series No. 115 in 1996 [1]. This publication was the culmination of efforts over past decades towards harmonization of radiation protection and safety standards internationally, and was jointly sponsored by the Food and Agriculture Organization of the United Nations (FAO), the International Atomic Energy Agency (IAEA), the International Labour Organisation (ILO), the OECD Nuclear Energy Agency (OECD/NEA), the Pan American Health Organization (PAHO) and the World Health Organization (WHO). The purpose of the Standards is to establish basic requirements for protection against the risks associated with exposure to ionizing radiation and for the safety of radiation sources that may deliver such exposure (hereinafter called 'radiation safety'). The requirements are based on the principles set out in the Safety Fundamentals publications. The fundamentals on radiation protection are published as IAEA Safety Series No. 120 [2], and the legal and governmental aspects are contained in IAEA Safety Standards Series No. GS-R-1 [3].

The BSS can only be implemented through an effective radiation safety infrastructure that includes adequate laws and regulations, an efficient regulatory system, supporting experts and services, and a 'safety culture' shared by all those with responsibilities for protection, including both management and workers. The BSS cover the application of ionizing radiation for all practices and interventions and are, therefore, basic and general in nature. Users of radiation sources have to apply these basic requirements to their own particular practices. In this context, the preamble of the BSS states that:

> "The [regulatory body] may need to provide guidance on how certain regulatory requirements are to be fulfilled for various practices, for example in regulatory guideline documents."

The present report does not contain requirements other than those quoted from the BSS, and therefore the only mandatory statements in the 'shall' form are quotations from the BSS. Any additional material is in the 'should' form or simply in the present tense, the latter indicating how to comply with the BSS.

1.2. OBJECTIVE

The objective of this report is to assist regulatory bodies in preparing regulatory guidance on the proper and consistent application of the BSS by the legal persons responsible for the radiotherapy practice.[1] This report will therefore also be useful to licensees in meeting regulatory requirements. Separate reports have been prepared for diagnostic radiology and interventional procedures using X rays, and for nuclear medicine.

1.3. SCOPE

This report is applicable to all established uses of ionizing radiation sources employed in the practice of radiotherapy, to the facilities where the sources are located and to the individuals involved. The guidance covers occupational, public, medical, potential and emergency exposure situations.

New techniques utilizing radiation sources (e.g. treatment of restenosis, stereotactic radiotherapy and intensity modulation radiotherapy) are not specifically addressed in this report, although the general principles of protection and safety discussed here are also applicable. This report also does not discuss unsealed sources, which are covered in a separate IAEA publication on nuclear medicine.

---

[1] Radiotherapy (radiation oncology) is the branch of clinical medicine that uses ionizing radiation, either alone or in combination with other modalities, for the treatment of patients with malignancies or other diseases. It includes responsibility for the diagnosis, treatment, follow-up and supportive care of the cancer patient as an integral part of the multidisciplinary management of patients. In many countries this specialized area of medicine is at the present time recognized under the term 'radiation oncology'. However, in this publication, the double terminology 'radiotherapy/radiation oncology' is used since radiation oncology excludes non-oncological treatment of benign diseases, whereas radiotherapy may also be used for the treatment of non-malignant conditions.

2. PRINCIPAL REQUIREMENTS OF THE BSS

2.1. ADMINISTRATIVE REQUIREMENTS

2.1.1. Authorization of the practice

The BSS require that legal persons apply to the regulatory body for an authorization, which takes the form of a registration or a licence. The BSS further clarify that:

"Typical practices that are amenable to registration are those for which: (a) safety can largely be ensured by the design of the facilities and equipment; (b) the operating procedures are simple to follow; (c) the safety training requirements are minimal; and (d) there is a history of few problems with safety in operations. Registration is best suited to those practices for which operations do not vary significantly."

Given the complexity of a radiotherapy practice, the risks involved and the fact that its safety depends largely on human performance and training, the demonstration of safety requires a specific safety assessment and therefore authorization is amenable to licence rather than to registration.

Setting up a radiotherapy practice involves the construction of facilities which are difficult to modify at a later time. Regulatory bodies may choose a two stage process of authorization, i.e. to require an initial application to construct a facility before construction begins. A good way to implement the two stage process is for the regulatory body to obtain an almost complete picture in the initial application, i.e. a description of the facility's design and equipment [4]. The regulatory body may also wish to place conditions on the procurement of radiation sources (including import) so that sources can only be imported when a particular stage of construction has been completed and safe storage of the sources can be ensured. The sequence may be subdivided into various steps (acceptance tests, commissioning, clinical use) for which additional information may be required by the regulatory body as a condition for allowing continuation of the process or for inspections to be performed.

Substantial modifications of the radiotherapy facilities, sources and procedures may have safety implications, which need regulatory verification of compliance. The regulatory body may also require a specific application for this. The same is true for partial or total decommissioning of a radiotherapy facility.

Radioactive sources and associated equipment for radiotherapy that had not been used for a long time while awaiting disposal have been involved in severe accidents when not properly secured or when security diminished with time [5, 6]. To prevent such accidents, there can be a requirement to notify the regulatory body of the planned date for resuming operation or of the decommissioning and disposal of the sources and the security conditions for interim storage. Three months is a recommended period of time. The longer the time period the higher the risk that control of the sources may be lost, but too short a period may increase bureaucracy without having a significant impact on safety.

2.1.1.1. Renewal of authorization

The regulatory body may consider a requirement to renew authorizations periodically. The renewal periods should be based on safety criteria and be established by the regulatory body. The advantages of a renewal or revalidation approach are described in Ref. [4] and in Ref. [7], which also describes the factors influencing the frequency of revalidation, i.e. the inspection frequency, the safety records and the stability of the user's operation. Considering these factors, a suitable period for renewal of radiotherapy authorizations may be five years.

2.1.2. Personal accreditation for radiation protection and safety[2]

The BSS, in para. 2.30, require that:

"(a) all personnel on whom protection and safety depend be appropriately trained and qualified so that they understand their responsibilities and perform their duties with appropriate judgement and according to defined procedures".

In the practice of radiotherapy the following individuals[3] have responsibility for protection and safety by virtue of tasks involving decisions, operation or manipulation of sources or equipment used in radiotherapy:

---

[2] Regulations in a number of countries require a personal authorization as formal recognition of the holder's competence to do the job safely.

[3] In Europe, radiation oncologists are usually referred to as radiotherapists.

— Radiation oncologists;[4]
— Qualified experts in radiotherapy physics (medical physicists) and dosimetrists;[5]
— Other health professionals operating radiotherapy equipment or handling radioactive sources (radiotherapy technologists)[6];
— The radiation protection officer;[7]
— Staff for maintenance of radiotherapy equipment;
— Staff performing special tasks (type tests, long term stability checks, etc.).[8]

To comply with the above BSS requirements in relation to the above staff, evidence is required of education and training relevant to their duties in relation to protection and safety. Responsibility for the practice of radiotherapy requires accreditation by the professional body or an educational institution.

Training in radiation protection is necessary, but by no means sufficient, to practice radiotherapy. As a precondition, the proper qualifications and certification in the profession are indispensable; these are usually not defined by radiation protection regulations, nor are they granted by the regulatory

---

[4] The radiation oncologist is specialized in the use of ionizing radiation for the treatment of cancer. He/she needs to not only have knowledge of his/her specialty, but also be knowledgeable of other alternatives for the treatment of cancer (surgery, chemotherapy, hormone therapy, etc.). An essential aspect of the radiation oncologist's job is to decide which of the possible curative or palliative therapeutic procedures should be used for each patient.

[5] The dosimetrist, under the supervision of the radiation therapy physicist, calculates the dose, plans the treatment and constructs treatment accessories. It is often the dosimetrist who periodically measures the absorbed dose rate from external sources and carries out the quality control tests designed by the radiation therapy physicist [5].

[6] In some countries, such as the USA, radiotherapy technologists are called radiation therapists, radiographers or radiotherapy technicians. The radiotherapy technologist or radiation therapist is responsible for operating the equipment and for positioning of the patient. He/she assists the radiation oncologist and the radiation therapy physicist in carrying out the treatment planned by them.

[7] Reference [8] indicates that in small institutions both functions (those of the qualified expert in radiotherapy physics and those of the radiation protection officer) may be fulfilled by the same individual, depending on their education and training. It should be borne in mind that the functions are different, and not that different persons are required to fulfil them.

[8] In some countries, long term stability checks and type tests may be performed only by persons with special permission from the regulatory body.

body. They are granted by academic institutions and by boards or societies. In the particular case of qualified experts, the BSS define them as:

> "An individual who, by virtue of certification by appropriate boards or societies, professional licences or academic qualifications and experience, is duly recognized as having expertise in a relevant field of specialization, e.g. medical physics".

For radiation oncologists, qualified experts in radiotherapy physics, dosimetrists, radiotherapy technologists and radiation protection officers, the typical documentary evidence indicated above, i.e. qualification credentials, consists of:

(a) A degree relevant to the profession, issued by the competent education and examining authorities, and accreditation issued by boards or societies required in the country to exercise the profession.
(b) A course in radiation protection for which the contents, methodology and teaching institution are approved by the regulatory body. This course may be integrated into the curriculum of the professional education under (a), provided that it meets the training criteria specified by the regulatory body.
(c) On the job training supervised by accredited professionals with experience, as required in the country, before working without supervision.

The courses and syllabus required in the professional education and training programme are generally defined by the departments of health and/or education in a country, together with the relevant professional bodies. It is acceptable for training criteria dealing with radiation protection for medical exposure, as specified by the regulatory body in consultation with the relevant professional bodies[9] (see BSS para. II.1(f)), to be incorporated into the professional education and training programme.

It may be appropriate and convenient for the regulatory body to recognize certain training centres and courses for their quality and suitability in connection with the radiation protection requirements. For example, it can identify: (a) radiation oncology departments that have been accredited as training centres for the profession (if any) and facilities; and (b) the syllabus

[9] In countries where a national professional body does not exist, a regional body or international professional organizations may be consulted for advice.

and qualifying bodies that are responsible for training and accreditation in radiation oncology, and recognize them for training in radiation protection as well. Such recognition can be formally conferred by a process of accreditation based on the training criteria referred to above.

Educated and trained professionals are an essential prerequisite for quality and safety in radiotherapy. However, the fact that educated professionals such as medical physicists leave their countries because of lack of recognition as health professionals is a major impediment to compliance with the requirements of the BSS. In this respect, the Action Plan on the Radiological Protection of Patients recognizes that in many countries the availability of medical physicists is limited by a lack of official recognition of these people as health professionals, and includes a measure to help solve this problem.

With regard to individuals engaged in maintaining radiotherapy equipment, the documentary evidence of competence consists of:

(1) Certification by the manufacturer of completion of a training programme on the type of authorized equipment (the certification should indicate the type of equipment and the parts of the equipment that the engineer or technician has been trained to repair or adjust, or the scope of the maintenance he/she is enabled to perform);
(2) A course on radiation protection for which the contents, methodology and teaching institution are approved by the regulatory body.

Personal accreditation or authorization may need to be renewed periodically. The regulatory body may provide guidance on qualification requirements in radiation protection for each category of job found in particular practices.

2.1.3. Authorization of other practices related to radiotherapy

Considering that the BSS require that the activities listed below be subject to authorization, regulatory bodies may require the licensee of a radiotherapy practice to contract any of the following services only to enterprises authorized by the regulatory body:

(a) Import, distribution, sale or transfer of radioactive sources;
(b) Installation and maintenance of radiotherapy equipment, including source change and decommissioning;
(c) Disposal of radioactive sources.

The requirements to carry out these practices should have been established by national regulations complemented by regulatory guidance documents.

2.1.4. Inspection

The BSS, in para. 1.10, require that:

> "The principal parties shall permit duly authorized representatives of the [regulatory body]...to inspect their protection and safety records and to carry out appropriate inspections of their authorized activities."

A sample list of items to be inspected in radiotherapy is provided in Appendix I.

2.2. RADIATION PROTECTION REQUIREMENTS

The radiation protection requirements on justification of a practice, dose limitation and optimization of protection, and the use of dose constraints (BSS paras 2.20–2.26) apply to radiotherapy. Table 1 summarizes the principles of radiation protection as applied to occupational and public exposure and to medical exposure. Dose limits do not apply to medical exposure and are not relevant for the control of potential exposures, nor are they relevant for decisions on whether and how to undertake an intervention. However, workers undertaking an intervention are subject to the relevant requirements of Appendix V of the BSS. Table 2 summarizes the values of dose limits. As indicated in Section 4, on occupational exposure in connection to pregnant workers:

"the employer of a female worker who has notified pregnancy shall adapt the working conditions in respect of occupational exposure so as to ensure that the embryo or foetus is afforded the same broad level of protection as required for members of the public" (BSS para. 1.17).

The term 'required dose to the target' from the BSS is considered to be the dose that the radiation oncologist decides to deliver. The therapeutic decision itself is a trade-off between two antagonistic objectives: to maximize the probability of local tumour control while keeping the probability and severity of complications to normal tissue as low as practicable. Optimized beam geometry and organ shielding and other means of improving dose distributions are commonly used in radiotherapy. In recent years it has been

TABLE 1. PRINCIPLES OF RADIATION PROTECTION AS APPLIED TO OCCUPATIONAL AND PUBLIC EXPOSURE AND TO THERAPEUTIC MEDICAL EXPOSURE

Principles of protection	
Application in general	Specific application to medical exposure
Justification of practices: The BSS require that "a practice that entails exposure to radiation should only be adopted if it yields sufficient benefit to the exposed individuals or to society to outweigh the radiation detriment."	Justification: By weighing the therapeutic benefits they produce against the radiation detriment they might cause, taking into account the benefits and risks of available alternative techniques that do not involve medical exposure.
Limitation of doses to individuals (for occupational and public exposure).	Dose limitation: Not applicable to medical exposure.
Optimization of protection: Providing the best available protection and safety measures under the prevailing circumstances, so that the magnitude and likelihood of exposures and the numbers of individuals exposed will be as low as reasonably achievable.	Optimization of protection: In therapeutic medical exposure, keeping exposure of normal tissue as low as reasonably achievable consistent with delivering the required dose to the planning target volume.[a]

[a] Planning target volume is a geometrical concept used in radiotherapy for planning treatment with consideration of the net effect of movements of the patient and of the tissues to be irradiated, variations in size and shape of the tissue, and variations in beam geometry such as beam size and beam direction.

possible to increase the dose to the target volume for some types of treatment, to achieve better local tumour control, keeping the normal tissue dose as low as before due to improved dose distribution.

The BSS, in para. 2.26, state that:

> "Except for medical exposure, the optimization of protection and safety measures associated with any particular source within a practice shall be subject to dose constraints".

Constraints are therefore not applicable to medical exposure of patients for their own diagnosis and treatment, but they apply to occupational and public exposure as well as for volunteers in biomedical research when the exposure is not to the benefit of the exposed person, and for comforters of patients. Dose constraints for occupational exposure in radiotherapy are dealt

TABLE 2. SUMMARY OF DOSE LIMITS[a]

Dose	Occupational exposure	Apprentices 16–18 years of age who are in training for employment and for students of 16–18 years	Public exposure
Effective dose	20 $mSv \cdot a^{-1}$ averaged over five consecutive years; 50 mSv in a single year	6 $mSv \cdot a^{-1}$	1 $mSv \cdot a^{-1}$ in special circumstances, an effective dose of up to 5 mSv in a single year, provided that the average dose over five consecutive years does not exceed 1 $mSv \cdot a^{-1}$
Equivalent dose to the lens of the eye	150 $mSv \cdot a^{-1}$	50 $mSv \cdot a^{-1}$	15 $mSv \cdot a^{-1}$
Equivalent dose to the extremities (hands and feet) or the skin[b]	500 $mSv \cdot a^{-1}$	150 $mSv \cdot a^{-1}$	Equivalent dose to the skin of 50 $mSv \cdot a^{-1}$

[a] See BSS Schedule II.

[b] According to the BSS, "The equivalent dose limits for the skin apply to the average dose over 1 cm^2 of the most highly irradiated area of the skin. Skin dose also contributes to the effective dose, this contribution being the average dose to the entire skin multiplied by the tissue-weighting factor for the skin."

with in Section 4.2. When protection is being optimized in the planning stage, a prospective assessment of individual doses is necessary and doses should be compared with the appropriate dose constraints.

A safety culture is to be inculcated that governs: attitudes and behaviour in relation to the protection and safety of all individuals and organizations dealing with sources of radiation; in-depth defensive measures which should be incorporated into the design and operating procedures to ensure that the safety objective is achieved despite potential failures in protection or safety measures; sound management and good engineering, QA, training and qualification of personnel, comprehensive safety assessments and attention to lessons learned from experience and research. Further information is given in Section 2.3.1.

2.3. MANAGERIAL REQUIREMENTS

2.3.1. Managerial commitment and policy statement

The BSS, in para. 2.28, establish that "A safety culture shall be fostered and maintained to encourage a questioning and learning attitude to protection and safety and to discourage complacency". To comply with this requirement, hospital management needs to be committed to an effective protection and safety policy, particularly at the senior level, and to demonstrable support for those persons with responsibility for radiation protection. The commitment can be demonstrated by a written policy that, while recognizing that the objective of the practice is the treatment and well-being of the patients, assigns the required level of importance to protection and safety. It is necessary to make this unambiguous statement known to hospital personnel by establishing a QA programme that provides for compliance with radiation protection requirements and fosters safety culture in the hospital.

2.3.2. Organization and responsibilities

The BSS, in paras 1.6 and 1.7, establish that:

"1.6. The principal parties having the main responsibilities for the application of the Standards shall be:

(a) registrants or licensees; and
(b) employers.

"1.7. Other parties shall have subsidiary responsibilities for the application of the Standards. These parties may include, as appropriate:

(a) suppliers;
(b) workers;
(c) radiation protection officers;
(d) medical practitioners;
(e) health professionals;
(f) qualified experts;
(g) Ethical Review Committees; and

(h) any other party to whom a principal party has delegated specific responsibilities."[10]

The licensee needs to assign clear subsidiary responsibilities to personnel (medical practitioners (radiation oncologists), qualified experts in radiotherapy physics, radiotherapy technologists, radiation protection officers and other health professionals) so that an adequate level of radiation protection of patients, workers and the public is ensured. The broad responsibilities of medical practitioners (radiation oncologists) and qualified experts in radiotherapy physics with regard to the BSS requirements on medical exposure are dealt with in Section 5.

According to the BSS, para. 1.9, it is also the responsibility of the licensee:

> "to establish protection and safety objectives in conformity with the relevant requirements of the Standards [to] develop a radiation protection and safety programme [and to] develop, implement and document a protection and safety programme commensurate with the nature and extent of the risks associated with the practices",

in this case radiotherapy. For the programme to be effective the licensee needs to provide for its implementation, including the necessary resources and arrangements to facilitate cooperation between all relevant parties.

An effective way to ensure compliance with the programme is to appoint a committee for radiation protection with the function of supervising safe operation and compliance with regulatory requirements. Since a representative of management (the licensee) is usually a member of the radiation protection committee, communication with that person may be the most appropriate. The members of the committee typically include an administrator representing the management (the licensee), the chief radiation oncologist, a qualified expert in radiotherapy physics (medical physicist), the radiation protection officer, a radiotherapy technologist, possibly a brachytherapy nurse and a maintenance engineer. A suggested list of items for the programme is given in Appendix I.

For the day to day oversight of the radiation protection programme, a radiation protection officer is needed, who should report to the committee. The licensee should provide this person with the time and resources required to supervise the programme, as well as the authority to communicate not only

[10] The BSS refer to medical practitioners in general which, in the case of a radiotherapy practice, are radiation oncologists.

with the committee on a periodic basis, but also directly with the licensee in the case of breaches of compliance which may compromise safety.

2.3.3. Quality assurance

The World Health Organization [9] has defined QA in radiotherapy as:

"[A]ll those procedures that ensure consistency of the medical prescription and the safe fulfillment of that prescription as regards to the target volume, together with minimal dose to normal tissue, minimal exposure of personnel, and adequate patient monitoring aimed at determining the end result of the treatment".

Similarly, IAEA-TECDOC-1040 [10] states that:

"[Q]uality assurance consists of procedures that ensure a consistent and safe fulfillment of the dose prescription to the target volume with minimal dose to normal tissues and minimal exposure to personnel and the public. It involves both clinical and physics aspects. The main areas will include clinical policies, treatment planning and delivery, a quality control programme for machine and equipment performance, maintenance programmes and investigative procedures for accidental medical exposures. The establishment of such a comprehensive quality assurance programme shall be in accordance with the Standards and the guidelines given by WHO" [9].

The BSS, in para. 2.29, require the licensee to have quality assurance programmes that provide:

"[A]dequate assurance that the specified requirements relating to protection and safety are satisfied [and] quality control mechanisms and procedures for reviewing and assessing the overall effectiveness of protection and safety measures."

It is an extensive and still growing practice for hospitals, especially radiotherapy departments, to implement a quality assurance system for medical care throughout the treatment period, i.e. covering the overall radiotherapy practice. This system involves a quality assurance committee.

The radiation protection committee, which deals with occupational, public and medical exposure, and the quality assurance committee, which deals with ensuring consistency of the medical prescription and the safe fulfillment of that prescription, have overlapping functions, especially with regard to the BSS requirements on radiation protection for medical exposure. Membership of both committees may be identical: an administrator representing management, the chief radiation oncologist, a qualified expert (medical physicist), a radiotherapy technnician, possibly a brachytherapy nurse and a maintenance engineer. Provision is needed for harmonizing the work of both committees.

The programme covers the entire process from the initial decision to adopt a particular procedure to the interpretation and recording of results, and includes ongoing auditing, both internal and external, as a systematic control methodology. The maintenance of records is an important part of QA, as is continuous quality improvement (CQI). It implies a commitment of the staff to strive for continuous improvement in the use of radiation sources in therapy, based on new information learned from their QA programme and new techniques developed by the radiotherapy community at large. Feedback from operational experience and lessons learned from accidents or near misses can help identify potential problems and correct deficiencies in radiation protection, and therefore needs to be systematically included in the CQI.

2.3.4. Human factors

The BSS, in para. 2.30, establish that:

"Provision shall be made for reducing as far as practicable the contribution of human error to accidents and other events that could give rise to exposures, by ensuring that:

(a) all personnel on whom protection and safety depend be appropriately trained and qualified so that they understand their responsibilities and perform their duties with appropriate judgement and according to defined procedures."

2.3.5. Staffing

To comply with the above requirement, the licensee has to appoint a sufficient number of professionals, with personal accreditation for the tasks described in Section 2.1.2, to ensure that all activities relevant to protection and safety are carried out in accordance with regulations and the radiation protection programme. It is important to keep the number of persons under review, especially as workload increases or new techniques and new equipment are incorporated.

2.3.6. Education and training

A number of requirements in the BSS refer to the availability of qualified personnel. The BSS, in para. 2.14, establish that:

“The legal person responsible for a source to be used for medical exposure shall include in the application for authorization:

(a) the qualifications in radiation protection of the medical practitioners [in this case radiation oncologists] who are to be so designated by name in the registration or licence; or
(b) a statement that only medical practitioners with the qualifications in radiation protection specified in the relevant regulations or to be specified in the registration or licence will be permitted to prescribe medical exposure by means of the authorized source.”

Paragraph 2.31 requires that “qualified experts shall be identified and made available”, and in particular, para. II.1(d) requires that “for therapeutic uses of radiation (including teletherapy and brachytherapy), the calibration, dosimetry and quality assurance requirements of the Standards be conducted by or under the supervision of a qualified expert in radiotherapy physics;” and para. II.1(c) requires that “medical and paramedical personnel [with appropriate training] be available as needed.”

Investment in radiotherapy equipment without concomitant investment in training can lead to a dangerous situation for patients and staff. The licensee has to ensure that only staff with the credentials specified in Section 2.1.2 fill related positions and that they are aware of:

(a) The conditions and limitations of the licence;
(b) The institutional radiation protection policies and procedures (including practice drills);
(c) Their own individual (subsidiary) responsibilities;
(d) The use and operation of equipment;
(e) The local QA programme and quality control procedures, which should be in an accessible manual;
(f) Review of incidents and accidental exposures;
(g) Instructions provided to patients and caregivers.

The professional education and the training to obtain the necessary qualifications need to have been completed before commencement of duties and then continued as part of professional development and as required by the regulatory body. Furthermore, staff training or upgrading may be required whenever significant changes occur in duties, regulations, the terms of the licence or radiation safety procedures.

It is important that the licensee establish a policy that encourages and provides a continuing professional development programme with the aim of improving staff skills, maintaining familiarity with current practices and fostering a safety culture throughout the institution. Such training and development schemes can be accomplished through informal meetings of the radiotherapy department, seminars, accredited continuing education programmes or other means.

In addition to the staff needing accreditation, the following staff needs to be provided with specific instructions on radiation protection:

(1) Brachytherapy nurses;
(2) Staff who are not employed by the radiotherapy practice but need to enter controlled areas;
(3) Staff who transport radioactive materials or patients with implants within the institution.

Personnel with duties in the vicinity of radioactive sources used in radiotherapy shall be informed of the radiation hazard, details of the specific uses, and the radiation protection programme. The licensee needs to keep the initial and periodic instruction of personnel documented as part of its records.

3. SAFETY OF SOURCES

Defence in depth is defined in the glossary of the BSS as "the application of more than a single protective measure for a given safety objective such that the objective is achieved even if one protective measure fails". A single equipment fault or human error should, therefore, not result in an accident. The BSS establish the following requirement for defence in depth (BSS para. 2.35):

"A multilayer (defence in depth) system of provisions for protection and safety commensurate with the magnitude and likelihood of the potential exposures involved shall be applied to sources such that a failure at one layer is compensated for or corrected by subsequent layers, for the purposes of:

(a) preventing accidents that may cause exposure;
(b) mitigating the consequences of any such accident that does occur; and
(c) restoring sources to safe conditions after any such accident."

3.1. SAFETY IN THE DESIGN OF RADIATION SOURCES AND EQUIPMENT

For sources used in medical exposure, the BSS establish that:

"II.11. The requirements for the safety of sources specified in other parts of the Standards shall also apply to sources used in medical exposure, where relevant and, in particular, equipment used in medical exposure shall be so designed that:

(a) failure of a single component of the system be promptly detectable so that any unplanned medical exposure of patients is minimized; and
(b) the incidence of human error in the delivery of unplanned medical exposure be minimized.

"II.12. Registrants and licensees shall:

(a) taking into account information provided by suppliers, identify possible equipment failures and human errors that could result in unplanned medical exposures;
(b) take all reasonable measures to prevent failures and errors, including the selection of suitably qualified personnel, the establishment of adequate procedures for the calibration, quality assurance and operation of diagnostic and therapeutic equipment, and the provision to personnel of appropriate training and periodic retraining in the procedures, including protection and safety aspects;
(c) take all reasonable measures to minimize the consequences of failures and errors that may occur; and
(d) develop appropriate contingency plans for responding to events that may occur, display plans prominently, and periodically conduct practice drills."

3.1.1. Equipment

Radiation sources, including radioactive material, equipment and accessories, should be purchased only from authorized suppliers and should have a valid type test.[11] Procedures for the purchase, installation, acceptance, commissioning, use, maintenance and quality control of such material should be

[11] Certain tests, termed 'type tests', refer to a type or brand of equipment and do not need to be repeated for all pieces of equipment. Individual tests refer to quality control of every piece of equipment.

developed with the involvement of qualified experts and the quality assurance/ radiation protection committee. According to Appendix II of the BSS:

"II.13. Registrants and licensees, in specific co-operation with suppliers, shall ensure that, with regard to equipment consisting of radiation generators and that containing sealed sources used for medical exposures:

(a) whether imported into or manufactured in the country where it is used, the equipment conform to applicable standards of the International Electrotechnical Commission (IEC) and the ISO or to equivalent national standards;
(b) performance specifications and operating and maintenance instructions, including protection and safety instructions, be provided in a major world language understandable to the users and in compliance with the relevant IEC or ISO standards with regard to 'accompanying documents', and that this information be translated into local languages when appropriate;
(c) where practicable, the operating terminology (or its abbreviations) and operating values be displayed on operating consoles in a major world language acceptable to the user".

The international standards applicable to radiotherapy are:
IEC-60601-2-1, for medical electron accelerators [11];
IEC-60601-2-11, for gamma external beam therapy [12];
IEC-60601-2-17, for remote afterloading[12] brachytherapy [13];
IEC-601-2-8, for superficial therapy with X rays [14];
IEC-60601-2-29, for therapy simulators [15];
IEC-62083, for treatment planning systems [16];
IEC-60601-1-4, for computer controlled systems or programmable electrical medical systems (PESS) [17].

Evidence of compliance with IEC or equivalent national standards is required. For type tests, this evidence may be sufficiently provided by manufacturer's records with the results of the tests for the relevant equipment

[12] Afterloading brachytherapy is the technique by which applicators or guides are placed on the patient prior to placement of the radioactive source, permitting verification of correct positioning without exposure of the staff, as well as prompt loading and unloading of radioactive material. Manual afterloading is the technique by which the loading and unloading of the sources are carried out manually, while remote afterloading equipment allows the loading and unloading of the radioactive sources by remote control.

type and model. Type tests have to be supplemented by acceptance tests for each individual piece of equipment delivered. Acceptance test protocols should include the relevant safety tests described in the IEC standards and it is a good practice to specify this method in the purchasing conditions. More detailed guidance is provided in Ref. [10]. IEC standards provide for tests to be carried out by the manufacturer for a given type of equipment, and for 'site tests' to be done at the hospital on every individual piece of equipment. The IEC distinguishes three grades of tests:

(1) Grade A: This grade refers to an analysis of the equipment design related to an IEC safety requirement, which results in a written statement included in the technical description, regarding the working principles or constructional means by which the IEC requirement is fulfilled.
(2) Grade B: Visual inspection, or functional test or measurement. For this test grade the relevant IEC standard specifies test procedures (see, for example, IEC 60601-2-1 [11]). Grade B tests may include fault conditions which require interference with the circuitry or construction of the equipment.
(3) Grade C: Functional test or measurement, which may involve interference with circuitry or the construction of the equipment and should be performed by, or under the direct supervision of, the manufacturer or its agent.

Equipment design should permit interruption[13] of the irradiation from the control panel; after the interruption, resumption of irradiation should be possible only from the control panel. External beam therapy equipment containing radioactive sources and high dose rate brachytherapy (HDR) equipment should be provided with a device to return sources manually to the shielded position in case of an emergency. For gamma knife units it should be possible to close the shielding door manually.

Irradiation heads in external beam therapy equipment, source containers in brachytherapy and other devices containing radioactive sources should have a clear permanent sign indicating the existence of radioactive material (i.e. ISO symbol) [18]. In addition, when devices containing radioactive sources are outside the radiotherapy department they should be labelled with a danger warning recognizable to any member of the public. The ISO radiation symbol alone is not intended to be a warning signal of danger but only of the existence of radioactive material. Accidents involving members of the public have

[13] This does not apply to manual brachytherapy.

occurred although the ISO radiation symbol was present, but not recognized as indicating danger. A symbol recognizable as meaning danger and prompting a protective action is under consideration by the IAEA for dangerous radiation sources (for a definition of dangerous source, see Ref. [19]).[14]

Operating instructions are used not only by qualified experts, but more importantly by technologists and technicians who may not understand any major world language. In such cases the accompanying documents need to be translated into the local language. The translation requires a QA process to ensure proper understanding and avoid operating errors. The same applies to maintenance instructions in relation to maintenance and service engineers and technicians.

The BSS, in para. II.13, further require that:

"(d) radiation beam control mechanisms be provided, including devices that indicate clearly and in a fail-safe manner whether the beam is 'on' or 'off';
(e) as nearly as practicable, the exposure be limited to the area being…treated by using collimating devices aligned with the radiation beam;
(f) the radiation field within the examination or treatment area without any radiation beam modifiers (such as wedges [or multileaf collimators]) be as uniform as practicable and the non-uniformity be stated by the supplier; and
(g) exposure rates outside the examination or treatment area due to radiation leakage or scattering be kept as low as reasonably achievable."

The requirements for radiation generators and irradiation installations for radiotherapy are:

"II.15. Registrants and licensees, in specific co-operation with suppliers, shall ensure that:

(a) radiation generators and irradiation installations include provisions for selection, reliable indication and confirmation (when appropriate and to the extent feasible) of operational parameters such as type of radiation,

---

[14] Reference [19] defines a dangerous source as a source that could, if not under control, cause fatal or life threatening exposures or result in a permanent injury that reduces quality of life. This includes HDR brachytherapy sources and sources used in external beam therapy units. See Ref. [19] for activities of radionuclides that are considered dangerous from the point of view of loss of control.

indication of energy, beam modifiers (such as filters), treatment distance, field size, beam orientation and either treatment time or preset dose;

(b) irradiation installations using radioactive sources be fail-safe in the sense that the source will be automatically shielded in the event of an interruption of power and will remain shielded until the beam control mechanism is reactivated from the control panel;

(c) high energy radiotherapy equipment:

(i) have at least two independent 'fail to safety' systems for terminating the irradiation; and

(ii) be provided with safety interlocks or other means designed to prevent the clinical use of the machine in conditions other than those selected at the control panel;

(d) the design of safety interlocks be such that operation of the installation during maintenance procedures, if interlocks are bypassed, could be performed only under direct control of the maintenance personnel using appropriate devices, codes or keys".

The primary beam of an external beam unit should be directed only towards primary barriers with sufficient shielding. If a primary shielding is incorporated into the equipment, electrical or mechanical interlocks should be provided to avoid the beam being directed towards the secondary barriers when the shielding is not intercepting the beam.

3.1.2. Sealed sources

Requirement II.15(e) of the BSS establishes that:

"(e) radioactive sources for either teletherapy or brachytherapy shall be so constructed that they conform to the definition of a sealed source".

Sealed source is defined in the BSS glossary as "[r]adioactive material that is (a) permanently sealed in a capsule or (b) closely bounded and in a solid form. The capsule or material of a sealed source shall be strong enough to maintain leaktightness under the conditions of use and wear for which the source was designed, also under foreseeable mishaps." To meet the requirements of BSS II.15, sealed sources used for external beam therapy and brachytherapy should comply with ISO 2919 [20].

Applicators for brachytherapy should be manufactured specifically for the source or be compatible with it. Use of radioactive sources after the working lifetime recommended by the manufacturer should be continued only after leak testing and approval by the regulatory body. Where older teletherapy units containing ^{137}Cs and brachytherapy sources incorporating ^{226}Ra or old ^{137}Cs in preloaded applicators are still in use[15], efforts should be made to replace them as soon as practicable with afterloading sources which do not contain ^{226}Ra and their applicators [21]. Sources using beta emitters should be provided with low atomic number shielding to minimize bremsstrahlung while they are in storage and preparation for use. Other types of source, such as liquid filled balloons for intravascular brachytherapy, require additional safety considerations but, as previously mentioned, are not specifically covered in this report.

3.1.3. Facilities and ancillary equipment

As a general rule, the design of the radiotherapy facility needs to make provisions for safety systems or devices associated with the equipment and room. This includes electrical wiring related to emergency 'off' switches, as well as safety interlocks and warning signals.

Methodology and data for shielding calculation are presented in Ref. [22]. The nominal design dose in occupied areas is derived by the process of constrained optimization, i.e. selecting a source related dose constraint, with the condition that the individual doses from all relevant sources be well below the dose limits for the persons occupying the area to be shielded. However, when using constraints for shielding calculations, consideration should be given to the remark made in International Commission on Radiation Protection (ICRP) Publication 33, para. 256 [21], that actual dose values to individuals are 1/10 (for equivalent dose) to 1/30 of dose values of effective dose[16] used as shielding design parameters. This is due to a number of conservative assumptions made in the calculation.

Typical conservative assumptions are: attenuation by the patient is usually not considered; maximum possible leakage radiation is assumed; workload, use and occupancy factors are overestimated; and the persons to be protected are permanently in the most exposed place of the adjacent room. It is therefore necessary to achieve a balanced decision and avoid accumulation of overly conservative measures that may go beyond optimization.

---

[15] Caesium-137 sources in powder form.

[16] Since ICRP Publication 33 preceded ICRP Publication 60, the quantities used were 'dose equivalent' and 'effective dose equivalent' rather than equivalent dose and effective dose. However, the point made that some of the assumptions may be too conservative is equally applicable to the quantities of 'equivalent dose' and 'effective dose'.

For radiation monitoring equipment, BSS para. II.15 requires that:

"(f) when appropriate, monitoring equipment be installed or be available to give warning of an unusual situation in the use of radiation generators and radionuclide therapy equipment."

Additional information on the design of radiotherapy facilities can be found in Ref. [10] and IEC Publication 61859 [23].

3.1.3.1. Manual brachytherapy

Typical safety features for the storage and preparation of sealed radioactive sources for manual brachytherapy are:

(a) The room should be used only for source storage and preparation by designated and trained personnel.
(b) The room should be provided with a locked door to control access and maintain source security (see Section 3).
(c) A radiation sign should be posted on the door.
(d) There should be shielded storage (a safe) for all sources, the outer surface of which should be made of fireproof materials. The safe should be located near the preparation workbench to reduce the exposure of personnel during handling and transfer of sources.
(e) The safe should have compartments for different source activities. Each compartment should be marked so as to permit immediate and easy identification of its contents from the outside with a minimum of exposure.
(f) The workbench should be provided with L block shielding with a lead glass viewing window [24].
(g) The source handling area should be well illuminated and a magnifying glass in a fixed mounting should be available for viewing, in order to handle sources efficiently and with a minimum of radiation exposure.
(h) Devices for handling sources, typically forceps, should be available. They should be as long as practicable, compatible with efficient source handling. A device should be provided for threading sources expeditiously with the fingers protected by distance.
(i) Sources should be readily identifiable by sight. When radioactive sources of the same appearance but of different activities are used, they should be distinguishable, e.g. by different coloured threads or beads [25].
(j) The working surface for source preparation should be smooth and seamless to avoid losing small sources such as ^{192}Ir wire fragments.

(k) The source storage and preparation laboratory should have a sink for cleansing of sources, provided with a filter or trap suitable for preventing loss of sources through the sewer.
(l) There should be a clear indication of the radiation level in terms of ambient dose equivalent. This should be achieved either by an area radiation monitor that should be visible on entering the room and during any handling of the unshielded sources, or a survey meter should be available and in use during source handling;
(m) Space should be available for secure storage to enable decay of short half-life sources such as ^{192}Ir;
(n) Hand carried transport containers should be provided with long handles and the lid of the container should be securely fastened to prevent tipping and dropping of sources during transport. Containers should bear the radiation symbol as well as a warning sign.
(o) Space should be available for source transport trolleys with source containers.

It is preferable that patients' rooms be single and adjacent to one another. Where this is not possible, appropriate shielding between patients is necessary. Shielding should be provided for nurses and visitors of brachytherapy patients, for which movable shields may be used within patients' rooms, especially in the case of manual brachytherapy.

Prior to each treatment, movable shields should be placed close to the patient's bed in such a way that exposure of the nurses caring for the patient is minimized. This is achieved by anticipating the nurse's tasks, positions and movements throughout the room. The treatment room should contain a shielded storage container (large enough to accept the applicators if necessary) and a remote handling tool (forceps) in the event of a dislodged source.

Sterilization facilities for preloaded applicators, if these are still being used until replacement by remote after loading applicators, is possible and should be available in the preparation or treatment rooms in order to ensure sufficient protection.

An area monitor should be placed at the entrance so as to detect when a source or a patient with a source is leaving the room or the controlled area. In order to ensure that after treatment no source remains within the patient, clothes or bed linen, or in the area a portable monitor should be available for monitoring these items.

3.1.3.2. Remote control brachytherapy and external beam therapy

External beam therapy and HDR brachytherapy should be carried out in a treatment room designed for that purpose within the radiotherapy department, while LDR remote control brachytherapy can be performed in the ward in the area in which manual brachytherapy is performed.

With regard to the treatment room for HDR brachytherapy, Ref. [10] states that "If the feasibility of sharing a shielded treatment room between an HDR unit and another currently used treatment machine is considered, it should be carefully evaluated. To avoid scheduling problems considerations should include the anticipated number of HDR procedures as well as the number of external beam treatments. This report recommends against this strategy in most instances."

Access to the irradiation room should be furnished with a visible signal indicating that the radiation source is 'on' or 'off'. A door interlock or other suitable means to prevent unauthorized access should be provided and a power fail-safe area radiation monitor should be visible upon entering the room. The mechanism should be capable of maintaining interruption of irradiation until the door is closed and locked and it has been verified that no person but the patient is inside the room. After an interruption, provided no operating parameters are changed or reselected, it should be possible to restart irradiation, but only from the equipment's control panel. One or more emergency off switches should be conveniently placed inside the treatment room to allow interruption of the irradiation from inside the room.

The control panel should be installed in such a way that the operator will have a total overview of the access to the irradiation room at all times. Adequate systems, devices or other means should be provided to allow the operator to have a clear and full view of the patient.

The systems for patient observation should be redundant and independent (e.g. closed circuit television or lead glass windows, depending on the type of treatment unit). Oral communication should be possible with the treatment rooms and patients using an intercom or other communication system. The presence of other staff in the area should be kept to the minimum necessary to avoid distraction to the operator.

Fire fighting means should be available in order to preserve the integrity of radioactive sources in the event of a fire. A radiation monitor and/or a portable survey instrument should be used to confirm the safe condition of the source.

3.2. SAFETY ASSOCIATED WITH ACCEPTANCE TESTS, COMMISSIONING AND OPERATION

Acceptance tests and commissioning should not be restricted to radiation emitting equipment or sources, but should also be conducted for any system that has implications for safety, such as treatment planning systems (TPSs). Insufficient understanding of TPSs at the commissioning stage and thereafter was involved in an accidental medical exposure [26].

3.2.1. Acceptance

After equipment installation, acceptance testing should be conducted in order to verify that the equipment conforms to technical specifications given by the manufacturer and to verify compliance with safety requirements from IEC standards. Acceptance tests should be performed, in the presence of personnel representing the user and personnel representing the manufacturer, by an individual or individuals acceptable to both parties.[17]

As discussed in Section 3.1, the tests to be included in the acceptance protocol should be specified in the purchasing conditions, and contracts should clearly establish the responsibility of suppliers for resolving non-conformity identified during acceptance testing. The grade B and C tests specified in the IEC standard (see Section 3.1) for a particular machine can be used as guidance for preparing the test protocol.

3.2.2. Commissioning

After acceptance and before starting operation, radiation sources and radiation beams are calibrated and commissioning is performed. These phases are critical to safety as shown in accidental exposures involving a large number of patients [27–29]. During commissioning the qualified expert in radiotherapy physics measures all data required for clinical use.

---

[17] Usually the equipment belongs to the supplier until the acceptance process has been completed. For this reason, in some countries a representative of the manufacturer carries out the tests in the presence of personnel representing the user (qualified expert in radiotherapy physics) to decide on acceptance.

3.2.3. Operation

Equipment should be operated in accordance with the technical documents, ensuring satisfactory operation at all times with respect to both the tasks to be accomplished and radiation safety. In particular, the manufacturer's operating manual and any additional procedures should be approved in accordance with the QA system (see Sections 2.3 and 3.1 for BSS requirements on equipment).

Sealed sources should be subject to leak tests prior to their first use and at regular intervals thereafter, in conformity with ISO 9978 [30]. Leak tests should be capable of detecting the presence of 0.2 kBq of removable contamination. For manual brachytherapy sources the typical method is the direct wet wipe test, while for external beam therapy and remote control brachytherapy the method to be used is the indirect wipe test of the nearest accessible surface. For ^{226}Ra sources, immersion or gas emanation tests are adequate; however, as indicated in Section 3.1.1, ^{226}Ra should be replaced by other radionuclides as soon as practicable. The sterilization process in brachytherapy should be appropriate for preventing damage to sources and applicators that could affect safety.

Periodic quality controls following formally established quality control protocols are necessary after the source has been installed or replaced, or after repairs or maintenance work that might alter the radiation output. A significant accidental exposure occurred because this was not done following a repair [27, 28, 31]. An independent audit of the calibration of the source should be carried out before clinical use of the source is started, e.g. the IAEA/WHO postal quality audit for dosimetry. The BSS requirements on QA for medical exposure are also provided in Section 5.

3.2.4. Maintenance

The licensee should ensure that adequate maintenance is done (preventative and corrective) and that inspections are carried out as necessary to ensure that radiation sources retain their design specifications for radiation protection and safety throughout their useful lives. This requires that the licensee establish the necessary arrangements and coordination with the manufacturer's representative before initial operation and on an ongoing basis.

The licensee should ensure that removal from and return to clinical service of radiotherapy equipment for maintenance or source exchange:

(a) Is documented and a record is kept;

(b) Where maintenance of the therapy equipment or treatment planning equipment may affect the accuracy of the physical or clinical dosimetry or the safe operation of the equipment, a qualified expert in radiotherapy physics assesses whether any specific tests or measurements are to be made and whether the equipment is operating satisfactorily before it is used to treat patients.

A contingency plan may need to be implemented when radiotherapy equipment is out of service for maintenance or source exchange, which is usually done by another company. In this case the licensee should provide the radiotherapy staff with written procedures outlining their involvement, if any, and the scope and a description of responsibilities for these actions.

3.2.5. Safe operation of external beam therapy

Safe operation of external beam treatment units requires procedures for area surveys, interlock checks, wipe tests and procedures for emergencies such as a source becoming stuck in the on or partially on position. Such procedures require that the necessary equipment be available, calibrated and in working order, including:

(a) A radiation monitor registering units in microsieverts or greater;
(b) Wipe test capabilities (for radioactive sources);
(c) Personal alarm dosimeters, especially for emergency intervention.

The procedures for the use of this equipment should recognize that some instruments will lock up in a high radiation field and give erroneous readings. Hence the procedure should require a three step process:

(1) Check the battery;
(2) Check the monitor response with a check source;
(3) Turn the instrument on and start monitoring from outside the room in which the source is located, i.e. from the lower to the higher dose rate areas.

The presence of other staff in the area of the control panel should be kept to the minimum necessary so as to avoid distraction to the operator.

3.2.6. Safe operation of brachytherapy

The source strength (usually in terms of the reference air kerma rate [32]) of each brachytherapy source should be determined individually before it is used on a patient, using a calibrated ionization chamber. The source documentation should be checked carefully. It is essential that the unit of activity used for source calibration be the same as the unit of activity used in the TPS. Some of the accidental exposures in brachytherapy have been caused by errors in the manufacturer's specification of the activity of one or several sources, and others because the unit of activity used at the hospital differed from the unit stated by the manufacturer [27, 28].

Low dose rate (LDR) and HDR sources have in common certain operating procedures for their safe use:

(a) Source inventories should be maintained, showing the location and current activity of each source at the facility as well as its unique identifier. This may be either a colour coded or an alphanumeric identifier.
(b) Sources should never be left on preparation surfaces. They have to be in storage, in transit or in use.
(c) Leak tests (using moist wipes) need to be performed and documented periodically; the tests should have sufficient sensitivity to detect the presence of 0.2 kBq [30] of removable contamination.
(d) For the HDR unit the wipe tests are only carried out on the afterloading drive assembly and transport containers since the source itself has too high a dose rate to allow this sort of test.
(e) Area surveys are to be performed periodically around the source storage facilities for LDR and HDR sources.
(f) The storage facilities are to be marked to indicate that they contain radioactive materials, and instructions given on how to contact the responsible radiation safety individual in the event of an emergency.
(g) The storage facilities are to be kept locked at all times.
(h) After every brachytherapy treatment the patient has to be monitored with a radiation survey meter to ensure that no activity remains in the patient unintentionally.

The following information should be posted in the case of LDR brachytherapy, both manual as well as remote controlled: identification of the patient, sources, date and time of insertion and removal, nursing required, time allowance for nurses and visitors, and concise instructions for unplanned source and applicator removal and for dealing with an emergency. A patient with a removable source in or on his or her body should not leave the room unless accompanied by a hospital attendant.

The licensee should ensure that all brachytherapy sources are removed from the patient, except in the case of permanent implants. The patient should be monitored with a portable detector to ensure that no source remains in or on their person. Linen, dressings, clothing and equipment should be kept within the room where the removal of sources takes place until all sources are accounted for and should be monitored with a radiation detector, as should rubbish bins, soiled dressing bins and laundry baskets. Mobile containers and portable equipment containing radioactive sources should be moved to storage or to a secure place when not in use.

3.2.6.1. Safe operation of manual brachytherapy

After verification of the source strength, the source or source holder should be marked with unique identifiers (for example, a pre-established colour), to facilitate visual recognition and prevent the possibility of confusion between different sources. Containers utilized for transport of radioactive sources need to conform with the requirements established in the IAEA's Regulations for the Safe Transport of Radioactive Material [33].

The movements of the sources from the time they leave the safe until their return should be documented, with the signature of the person responsible for the move (using forms or a log book). A person should be assigned to be in charge of accountability for the sources. This person should keep a record of the source order and of issuance from and return to the safe, with signatures (see the requirements for source security below).

The sources are to be inspected visually for possible damage after each use by means of magnifying viewers and a leaded viewing window in a shielded work area. A diagram at the source storage safe, which has to show the exact location of each source within the safe, aids in reducing the time it takes to locate and identify a source. Sources should only be handled with long forceps or tongs, never directly with the fingers. A mobile shielded container is needed for transport of sources and the shortest route possible should be used.

Sources which come into direct contact with body tissues require cleaning and possible sterilization after each use, which can subject the sources to possible damage from heat, abrasion, chemical attack and mechanical stresses. Therefore these sources must be inspected after every use. The work surfaces should be easy to clean and brightly lit to make it easy to find dropped sources. As stated in Section 3.1, a filter or a trap should be used in drains to prevent loss of sources to the sewage during cleaning. A portable detector should be used to ensure that no source remains in or on the patient.

Precautions to be observed during the cutting and handling of ^{192}Ir wires should include ensuring that:

(a) Appropriate tools and equipment such as forceps, cutting devices, magnifying glasses and good illumination of the work surface are available and used and that if ^{192}Ir wires are cut off for immediate use a container to hold cut lengths is provided and labelled;
(b) Radioactive waste is collected and stored in adequate containers;
(c) Surfaces and tools are properly decontaminated.

3.2.6.2. Safe operation of remote control afterloading brachytherapy

The QC of the afterloader should include tests to be performed at the beginning of each treatment day. The couplings and transfer tubes need to be checked (for HDR this has to be done before each treatment) to ensure that there is nothing to prevent the source from moving as required.

Remote afterloading equipment requires specific emergency procedures, which are especially critical for HDR brachytherapy. These procedures are dealt with in Section 7.3.

3.3. SECURITY OF SOURCES

The BSS, in para. 2.34, require that:

"Sources shall be kept secure so as to prevent theft or damage and to prevent any unauthorized legal person from carrying out any of the actions specified in..., by ensuring that:

(a) control of a source not be relinquished without compliance with all relevant requirements specified in the registration or licence and without immediate communication to the [regulatory body]..., of information regarding any decontrolled, lost, stolen or missing source;
(b) a source not be transferred unless the receiver possesses a valid authorization; and

(c) a periodic inventory of movable sources be conducted at appropriate intervals to confirm that they are in their assigned locations and are secure." [18]

The objective of source security is to ensure continuity in the control and accountability of each source at all times in order to meet BSS requirement 2.28. Specific provisions are required for avoiding loss of control in the following situations:

(1) Storage of sources before installation;
(2) Temporary or permanent cessation of use;
(3) Storage after decommissioning while awaiting a decision on source return or disposal;
(4) Brachytherapy sources remaining in the patient, clothes, bed linen or treatment area.

To comply with these requirements, the licensee needs to develop procedures to ensure the safe exchange and movement of radioactive sources within the institution and establish controls to prevent theft, loss, unauthorized withdrawal or damage of sources, or entrance of unauthorized personnel to the controlled areas.

The licensee also needs to ensure that the number of sources in a container is checked when they are being removed and returned and that a physical inventory of all sealed sources is carried out to confirm that they are present and secure in their assigned locations. The licensee should maintain a source movement log with a record indicating the date of removal, the name of the patient and the return of the source.

Radiotherapy equipment should be provided with safety systems capable of preventing its use by unauthorized personnel. A key should be required to energize the system, access to which needs to be restricted to authorized staff. Any loss of a source needs to be reported immediately to the radiation protection officer, who should report it to the radiation protection committee and to the regulatory body. All linen, dressing, clothing, equipment and rubbish containers should be kept within the brachytherapy patient's room until checks have been made and it has been documented that no sources are attached to them.

---

[18] Legal person is defined in the BSS as "any organization, corporation, partnership, firm, association, trust, estate, public or private institution, group, political or administrative entity or other persons designated in accordance with national legislation, who or which has responsibility and authority for any action taken under these Standards".

To illustrate the kinds of accident that can be caused by insufficient security in dealing with radiotherapy sources (nearly abandoned sources), case histories of accidents resulting from breaches in security are given in Appendix IV and more details can be found in Refs [5, 6, 27, 28]. From the point of view of loss of source control and potential exposure of members of the public, the IAEA's categorization of sources [34] assigns the first category (highest level of risk) to radioactive sources used for external beam radiotherapy and the second level to high and medium dose rate brachytherapy, while the LDR brachytherapy sources are assigned level four.

4. OCCUPATIONAL EXPOSURE

Detailed requirements for protection against occupational exposure are given in the BSS and recommendations on how to meet these requirements are given in Refs [35, 36]. This section summarizes those most relevant to radiotherapy.

4.1. RESPONSIBILITIES AND CONDITIONS OF SERVICE

The BSS require that:

"I.1. Registrants and licensees and employers of workers who are engaged in activities involving normal exposures or potential exposure shall be responsible:

(a) for the protection of workers from occupational exposure; and
(b) for compliance with any other relevant requirements of the Standards.

"I.2. Employers who are also registrants or licensees shall have the responsibilities of both employers and registrants or licensees."

The parties responsible for occupational exposure are therefore not only registrants and licensees but also employers. Registrants, licensees and employers of workers are responsible for ensuring that exposure is limited, protection and safety are optimized and appropriate radiological protection programmes are set up and implemented. The BSS further require that

"I.9. Employers, registrants and licensees shall facilitate compliance by workers with the requirements of the Standards."

The BSS also establish the subsidiary responsibilities of workers:

"I.10. Workers shall:

(a) follow any applicable rules and procedures for protection and safety specified by the employer, registrant or licensee;
(b) use properly the monitoring devices and the protective equipment and clothing provided;
(c) co-operate with the employer, registrant or licensee with respect to protection and safety and the operation of radiological health surveillance and dose assessment programmes;
(d) provide to the employer, registrant or licensee such information on their past and current work as is relevant to ensure effective and comprehensive protection and safety for themselves and others;
(e) abstain from any wilful action that could put themselves or others in situations that contravene the requirements of the Standards; and
(f) accept such information, instruction and training concerning protection and safety as will enable them to conduct their work in accordance with the requirements of the Standards."

Workers are also responsible for providing feedback to management.[19] The BSS require that "If for any reason a worker is able to identify circumstances that could adversely affect compliance with the Standards, the workers shall as soon as feasible report such circumstances to the employer, registrant or licensee", and also prescribe that management "shall record any report received from a worker that identifies circumstances which could affect compliance with the Standards, and shall take appropriate action."

In some cases the employer, registrant and licensee are the same legal person, but in other cases they may be different people. For example, the employer of a maintenance engineer for radiotherapy equipment may be the maintenance company, and these itinerant engineers may work in many radiotherapy departments, each one under a different licensee. There is a need for cooperation of the employers, the workers and the managements of hospitals. The BSS require that

---

[19] The responsibilities are placed on the management of the organizations of registrants, licensees or employers. For simplicity, Ref. [35] uses the word 'management' to denote registrants, licensees or employers.

"I.30. If workers are engaged in work that involves or could involve a source that is not under the control of their employer, the registrant or licensee responsible for the source and the employer shall co-operate by the exchange of information and otherwise as necessary to facilitate proper protective measures and safety provisions."

The organizational structure should reflect the assignment of responsibilities and the commitment of the organization to protection and safety. The management structure should facilitate cooperation between the various individuals involved. The radiation protection programme should be designated in such a way that the relevant information is provided to the individuals in charge of the various aspects of the work [35].

A self-employed person is regarded as having the duties of both an employer and a worker, as specified in the BSS definition of 'worker' (see Definitions). This situation is very much applicable to radiotherapy, because in some private radiotherapy departments the head of the department is self-employed and can be an occupationally exposed worker.

4.2. USE OF DOSE CONSTRAINTS IN RADIOTHERAPY

Dose constraints can be used for optimizing protection in the planning stage for each radiation source. Anticipated individual doses should be compared with the appropriate dose constraints, and only protective measures that predict doses below dose constraints should be chosen. The BSS definition of dose constraint is: "For occupational exposures, dose constraint is a source related value of individual dose used to limit the range of options considered in the process of optimization." Dose constraints are not intended to be applied retroactively to check compliance with protection requirements but to assess individual doses at the design and planning stages.

Since dose constraints are source related, the source to which they are related needs to be specified, e.g. when choosing source related dose constraints for the sources involved in a radiotherapy facility, the fact that medical and paramedical staff may work in more than one hospital and be exposed to the sources from more than one radiotherapy department has to be taken into consideration (for example in one hospital in the morning and another in the evening). Further discussion of dose constraints can be found in paras 4.17–4.21 of Ref. [35].

4.3. INVESTIGATION LEVELS FOR STAFF EXPOSURE IN RADIOTHERAPY

The BSS Glossary defines investigation level as "The value of a quantity such as equivalent dose, intake, or contamination per unit area or volume at or above which an investigation should be conducted."

Investigation levels are a tool used to provide a 'warning' of the need to review procedures and performance, investigate what is not working as expected and take timely corrective action. In radiotherapy, a suitable unit for use as the investigation level is the monthly effective dose itself, but the dose to the hands can be used as a unit for the investigation level for staff in manual brachytherapy. In a radiotherapy department where different staff is dedicated to specific work or tasks, different investigation levels can be associated with the various tasks.

Following are examples of levels and their related tasks that are rarely exceeded and, therefore, may be suitable as investigation levels: for persons working only with accelerators or remote control brachytherapy, a monthly investigation level of 0.4 mSv effective dose; for staff working with ^{60}Co external beam therapy, brachytherapy nurses, and persons inserting and removing manual brachytherapy sources, a monthly investigation level of 0.5 mSv effective dose may be used.

4.4. PREGNANT WORKERS

The BSS establish that:

"I.16. A female worker should, on becoming aware that she is pregnant, notify the employer in order that her working conditions may be modified if necessary.

"I.17. The notification of pregnancy shall not be considered a reason to exclude a female worker from work; however, the employer of a female worker who has notified pregnancy shall adapt the working conditions in respect of occupational exposure so as to ensure that the embryo or foetus is afforded the same broad level of protection as required for members of the public."

Limitation of the dose to the conceptus does not mean that it is necessary for pregnant women to avoid work with radiation but it does imply a necessity for the employer to carefully review the exposure conditions with regard to both normal exposure and potential exposure. For example, the dose to the foetus for workers involved in manual brachytherapy, under normal conditions, may reach the dose limit for members of the public established in the BSS (see Section 2).

Special consideration should be given to assigning a pregnant woman duties in which accidents are very unlikely and avoiding her intervention in an emergency such as those described in Section 7, for example work with a cobalt unit or an HDR brachytherapy unit. Counselling for pregnant workers should be available as stated in Section 4.11.

4.5. CLASSIFICATION OF AREAS

Relevant areas of a practice can be classified as 'controlled' or 'supervised' (BSS paras I.21–I.25). A controlled area is defined as an area in which specific protection measures and safety provisions are needed to control normal exposure and to prevent potential exposure (see Definitions).

In radiotherapy practices, areas requiring specific protection measures (controlled areas) include, at least, all irradiation rooms for external beam therapy and remote afterloading brachytherapy, operating rooms during brachytherapy procedures using real sources, brachytherapy patient rooms, radioactive source storage and handling areas. It is preferable to define controlled areas by physical boundaries like walls or other physical barriers marked or identified with 'radiation area' signs.

A frequently asked question is whether the area around the control panel for external beam therapy should be a controlled or a supervised area. From the exposure point of view, under normal conditions a controlled area may not be strictly necessary as the shielding can be designed so that exposure is sufficiently low. However the area may still have to designated as a controlled area to restrict access and avoid distraction of the operator, which may lead to accidental medical exposure to patients..

A supervised area is any area not already designated as a controlled area but where occupational exposure conditions need to be reviewed, even though specific protection measures and safety provisions are not normally needed. Supervised areas may include the areas surrounding brachytherapy patients' rooms or around radioactive source storage and handling areas. Persons in controlled or supervised areas should be afforded the same level of protection as members of the public.

4.6. LOCAL RULES AND SUPERVISION

The BSS require that:

"I.26. Employers, registrants and licensees shall, in consultation with workers, through their representatives if appropriate:

(a) establish in writing such local rules and procedures as are necessary to ensure adequate levels of protection and safety for workers and other persons;
(b) include in the local rules and procedures the values of any relevant investigation level or authorized level, and the procedure to be followed in the event that any such value is exceeded;
(c) make the local rules and procedures and the protective measures and safety provisions known to those workers to whom they apply and to other persons who may be affected by them;
(d) ensure that any work involving occupational exposure be adequately supervised and take all reasonable steps to ensure that the rules, procedures, protective measures and safety provisions be observed; and
(e) when required by the Regulatory Authority, designate a radiation protection officer.

"I.27. Employers, in co-operation with registrants and licensees, shall:

(a) provide to all workers adequate information on the health risks due to their occupational exposure, whether normal exposure or potential exposure, adequate instruction and training on protection and safety, and adequate information on the significance for protection and safety of their actions;
(b) provide to female workers who are liable to enter controlled areas or supervised areas appropriate information on:
    (i) the risk to the embryo or foetus due to exposure of a pregnant woman;
    (ii) the importance for a female worker of notifying her employer as soon as she suspects that she is pregnant; and
    (iii) the risk to an infant ingesting radioactive substances by breast feeding;
(c) provide to those workers who could be affected by an emergency plan appropriate information, instruction and training; and
(d) keep records of the training provided to individual workers."

More specific rules and procedures for operation of external beam therapy and for brachytherapy are given in Section 3.2.

4.7. PROTECTIVE EQUIPMENT AND TOOLS

According to BSS, para. I.28, employers and licensees "shall ensure that...workers be provided with suitable and adequate personal protective equipment". In the case of radiotherapy, typical protective equipment is a shielded L block on the workbench and lead glass, as well as devices for handling sources, which are described in more detail in Section 3.1.3.

4.8. INDIVIDUAL MONITORING AND EXPOSURE ASSESSMENT

The BSS, in paras I.32–I.34, establish that:

"I.32. The employer of any worker, as well as self-employed individuals, and the registrants and licensees shall be responsible for arranging for the assessment of the occupational exposure of workers, on the basis of individual monitoring where appropriate, and shall ensure that adequate arrangements be made with appropriate dosimetry services under an adequate quality assurance programme.

"I.33. For any worker who is normally employed in a controlled area, or who occasionally works in a controlled area and may receive significant occupational exposure, individual monitoring shall be undertaken where appropriate, adequate and feasible. In cases where individual monitoring is inappropriate, inadequate or not feasible, the occupational exposure of the worker shall be assessed on the basis of the results of monitoring of the workplace and on information on the locations and durations of exposure of the worker.

"I.34. For any worker who is regularly employed in a supervised area or who enters a controlled area only occasionally, individual monitoring shall not be required but the occupational exposure of the worker shall be assessed. This assessment shall be on the basis of the results of monitoring of the workplace or individual monitoring."

The purpose of monitoring and dose assessment is, inter alia, to provide information on the actual exposure of workers and confirmation of good work practices. It contributes to reassurance and motivation. The BSS require individual monitoring for any worker who is normally employed in a controlled area and may receive significant occupational exposure.

Those most likely to require individual monitoring are radiation oncologists, qualified experts in radiotherapy physics, the radiation protection officer, radiotherapy technologists, source handlers, maintenance staff and any nursing or other staff who must spend time with patients who contain sources.

Monitoring includes more than just measuring. It includes interpretation and assessment. Individual external doses can be assessed by using individual monitoring devices such as thermoluminescent dosimeters or film badges, which are usually worn on the front of the upper torso (in most radiotherapy procedures the whole body is assumed to be fairly uniformly exposed). The operational dosimetric quantity required in the BSS [36] is the personal dose equivalent Hp(d).

For weakly penetrating and strongly penetrating radiation, the recommended depths are 0.07 mm and 10 mm, respectively. Radiation used in radiotherapy is usually strongly penetrating and therefore $d = 10$ mm, except in the case of use of beta sources for brachytherapy. Other depths may be appropriate in particular cases, for example 3 mm for the lens of the eye, if the dose to the eye is higher than for the rest of the body and therefore requires specific assessment. This is generally not the case in radiotherapy, where the handling of the sources for preparation and insertion should be done with the face protected by a workbench provided with L block shielding with a lead glass viewing window. When the possibility of substantial exposure to the hands exists, such as in the handling of brachytherapy sources, extremity dosimeters may need to be worn (if this is compatible with clinical practice).

Delays in the evaluation of a dosimeter can result in fading of the stored information. If an individual's dosimeter is lost, the dose the individual is likely to have received must be assessed and added to the worker's dose record. Often the most reliable method of estimating an individual's dose is to use his or her recent dose history, provided that nothing unusual has occurred during that period.

The use of additional operational dosimeters, such as electronic dosimeters, is also to be recommended for use in radiotherapy, as these devices can give the worker an instant indication of both the cumulative and the current dose rate and allow presetting of an alarm [35, 36]. In cases where occupational exposure comes from workplace monitoring (see BSS para. 1.34), the effective dose can be inferred from the ambient dose equivalent H*(10). Reference [37] provides conversion coefficients from ambient dose equivalent to effective dose for different types of radiation and energies. The conversion coefficients for photons are close to unity, except for very low energy such as the energy of scattered photons from a low kilovolt X ray beam. Individual monitoring devices should be calibrated and this calibration be traceable to a standards dosimetry laboratory. For more detailed guidance see Refs [35, 36].

4.9. INVESTIGATION AND FOLLOW-UP

Employers and licensees are to conduct formal investigations, as required by the regulatory body, whenever:

(a) The individual annual effective dose exceeds investigation levels;
(b) Any of the operational parameters subject to periodic quality control are out of the normal range established for operational conditions;
(c) An equipment failure or error causes, or has the potential to cause, an accident (e.g. a teletherapy or remote afterloader source fails to return to the shielded position);
(d) Any other change or unusual circumstance causes an increase in dose exceeding dose limits or the operational restrictions imposed on the installation (e.g. significant change in workload or operating conditions of radiotherapy equipment).

The investigation is to be initiated as soon as possible following the event and a written report is to be prepared concerning its cause, including determination or verification of any doses received, corrective or mitigating actions, and instructions or recommendations to avoid recurrence. The report is to be submitted to the regulatory body and other concerned bodies as soon as possible.

4.10. MONITORING OF THE WORKPLACE

The BSS, in paras I.37–I.40, require licensees to develop programmes for monitoring the workplace:

"I.38. The nature and frequency of monitoring of workplaces shall:

(a) be sufficient to enable:
  (i) evaluation of the radiological conditions in all workplaces;
  (ii) exposure assessment in controlled areas and supervised areas; and
  (iii) review of the classification of controlled and supervised areas; and
(b) depend on the levels of ambient dose equivalent and activity concentration, including their expected fluctuations and the likelihood and magnitude of potential exposures.

"I.39. The programmes for monitoring of the workplace shall specify:

(a) the quantities to be measured;

(b) where and when the measurements are to be made and at what frequency;
(c) the most appropriate measurement methods and procedures; and
(d) reference levels and the actions to be taken if they are exceeded."

Initial monitoring is to be conducted immediately after the installation of new radiotherapy equipment and after the replacement of teletherapy sources and remote controlled brachytherapy sources. Initial monitoring includes measurements of radiation leakage from equipment within acceptance tests and area monitoring of habitable space around irradiation rooms.[20]

Exposure levels are monitored through the use of area monitors in teletherapy and high dose rate treatment rooms. The source storage and handling area is to be monitored with a survey meter immediately following the removal from or return to storage of brachytherapy sources.

Monitoring is to be done in association with brachytherapy procedures. Soon after implantation of the sources a survey of exposure rates in the vicinity of the patient is necessary. After removal of brachytherapy sources from a patient, a survey is to be carried out to confirm removal from the patient and return to shielding of all sources. The transport container should be surveyed before and after brachytherapy procedures. Packages containing radioactive sources are to be monitored on their receipt by the licensee.

All survey meters used for workplace monitoring need to be calibrated and this calibration needs to be traceable to a standards dosimetry laboratory. For more detailed guidance see Ref. [37].

4.11. PROTECTION OF WORKERS IN INTERVENTIONS (EMERGENCIES)

Section 7 identifies and provides a short description of emergency situations in radiotherapy, their prevention, preparation for them and their mitigation. Emergencies involve loss of radiotherapy sources and stuck sources in external beam therapy and remote control brachytherapy units. In addition, there have been a number of major off-site accidents due to loss of control of teletherapy sources which were not in use.

---

[20] Ambient dose equivalent H*(10) can be used to estimate the personal dose equivalent Hp(10) that would correspond to an individual staying in the same radiation field. Conversion coefficients are given in Ref. [36]. Hp(10) provides an estimate of the effective dose that avoids both underestimation and excessive overestimation.

Radiotherapy workers will be involved in the mitigation actions, especially in the case of on-site emergencies. Although dose limits for practices do not apply to interventions, the safety guide on occupational protection [35] points out that the exposure of workers in interventions cannot be considered an unexpected exposure but rather is deliberate and controlled, and the dose limits for workers should be assumed to apply unless there is an overriding reason not to apply them, such as the need to save life after an accident or to prevent catastrophic conditions.

In emergency situations, contingency plans based on the events identified by the safety assessment include allocation of responsibilities and provide for training of the relevant staff in executing the mitigation measures, which are expected to be periodically rehearsed. These actions are, therefore, deliberate and controlled and there is generally no overriding reason for not applying the occupational dose limits to these situations.

4.12. HEALTH SURVEILLANCE

Paragraph I.41 of the BSS states that: "Employers, registrants and licensees shall make arrangements for appropriate health surveillance in accordance with the rules established by the [regulatory body]." The primary purpose of health surveillance is to assess the initial and continuing fitness of employees for their intended tasks. Health surveillance programmes should be based on the general principles of occupational health. It should be rare for the radiation component of the working environment to significantly influence the decision about the fitness of a worker to undertake work with radiation, or to influence the general conditions of service (see para. 7.6 of Ref. [35]). No specific health surveillance related to exposure to ionizing radiation is necessary for staff involved in the operation of a radiotherapy practice. Only in the case of workers overexposed at doses much higher than the dose limits (e.g. 0.2–0.5 Sv or higher) would special investigations involving biological dosimetry and further extended diagnosis and medical treatment be necessary (see para. 7.18 of Ref. [35]).

Counselling should be available to workers (para. 7.14 of Ref. [35]) such as women who are or may be pregnant, individual workers who have or may have been exposed substantially in excess of dose limits, and workers who may be worried about their radiation exposure. This is particularly necessary for women who are or may be pregnant such as, for example, female technologists working in radiotherapy and nurses working in brachytherapy wards.

4.13. RECORDS

According to BSS para. I.44, employers and licensees “shall maintain and preserve exposure records for each worker”. The exposure records shall include information on the general nature of the work involving occupational exposure; doses and the data upon which the dose assessments have been based; when a worker is or has been occupationally exposed while in the employ of more than one employer; the dates of employment with each employer and the doses, exposures and intakes in each such employment; and records of any doses due to emergency interventions or accidents, which shall be distinguished from doses incurred during normal work. Employers and licensees are to provide for access by workers to information in their own exposure records, and give due care and attention to the maintenance of appropriate confidentiality of records.

5. MEDICAL EXPOSURE

The detailed requirements given in Appendix II of the BSS are applicable, in particular, to radiotherapy. In addition, Ref. [8] describes strategies to involve organizations outside the regulatory framework, such as professional bodies, whose cooperation is essential to ensure compliance with the BSS requirements for medical exposures. Examples that may illustrate this point include the adoption of protocols for calibration of radiotherapy units and reporting of accidental medical exposure.

As an overall remark, it is important to note that the principles, justification and optimization of protection requirements also apply to medical exposure but not the dose limitation (see Table 1). Further, dose constraints do not apply to exposure of patients as part of their own diagnosis and treatment but specific dose constraints are to be defined for comforters and for medical exposure of individuals exposed for medical research if these individuals do not benefit directly from the exposure.

5.1. RESPONSIBILITIES

With regard to responsibilities for medical exposure, the BSS require that:

"II.1. Registrants and licensees shall ensure that:

(a) no patient be administered a diagnostic or therapeutic medical exposure unless the exposure is prescribed by a medical practitioner;
(b) medical practitioners [in this case radiation oncologists] be assigned the primary task and obligation of ensuring overall patient protection and safety in the prescription of, and during the delivery of, medical exposure;
(c) medical and paramedical personnel be available as needed, and either be health professionals or have appropriate training adequately to discharge assigned tasks in the conduct of the diagnostic or therapeutic procedure that the medical practitioner prescribes;
(d) for therapeutic uses of radiation (including teletherapy and brachytherapy), the calibration, dosimetry and quality assurance requirements of the Standards be conducted by or under the supervision of a qualified expert in radiotherapy physics".

In addition, as stated in the BSS (see Section 2.3 of this publication), subsidiary parties with responsibilities for compliance with safety standards can also be workers, radiation protection officers, health professionals or any other party to whom a principal party has delegated specific responsibilities. Each individual should take actions within his or her area of responsibility, as established in the radiation protection programme, to prevent inappropriate exposures of patients.

All persons involved in the delivery of medical exposure should:

"(a) Follow the applicable rules and procedures for the protection and safety of patients, as specified by the licensee;
(b) Be aware that prescription of treatment and treatment plan need to be signed by the medical practitioner (radiation oncologist) prior to initiation of treatment."

Furthermore, the BSS require that the licensees "shall ensure that:

"(e) the exposure of individuals incurred knowingly while voluntarily helping (other than in their occupation) in the care, support or comfort of patients undergoing medical diagnosis or treatment be constrained as specified in Schedule II; and
(f) training criteria be specified or be subject to approval, as appropriate, by the [regulatory body] in consultation with relevant professional bodies.

....

"II.3. Medical practitioners [in this case radiation oncologists] shall promptly inform the registrant or licensee of any deficiencies or needs regarding compliance with the Standards with respect to protection and safety of patients and shall take such actions as may be appropriate to ensure the protection and safety of patients."

To comply with these requirements, it is indispensable that registrants and licensees establish an internal mechanism to ensure that medical exposure is prescribed by a medical practitioner (in this case the radiation oncologist), that the obligation for the overall patient protection is assigned to a radiation oncologist, that medical and paramedical staff be available, that qualified experts in radiotherapy physics carry out or supervise the calibration, dosimetry and QA, and that only staff with the necessary training be in charge of exposing patients for treatment.

5.2. JUSTIFICATION

Pursuant to BSS para. II.4, justification of medical exposure is required: "Medical exposures should be justified by weighing the diagnostic or therapeutic benefits they produce against the radiation detriment they might cause, taking into account the benefits and risks of available alternative techniques that do not involve medical exposure." In Ref. [39], the ICRP describes three levels of justification.

In the case of radiotherapy, the medical practitioner should consider the efficacy, benefits and risks of alternative treatment modalities, e.g. surgery and chemotherapy, either alone or in combination with radiation therapy.

The justification of therapeutic exposure of pregnant patients deserves specific consideration. The following is adapted from Ref. [38]. Cancer in pregnancy is relatively uncommon (about 0.1%) but it constitutes a major problem for both physicians and patients. As an example, in the USA alone about 4000 women per year are considered for radiotherapy during pregnancy. The ethical and risk/benefit issues for the patient in this setting are quite different from the use of most medical radiation where the patient has both the benefit and the risk. In radiotherapy of pregnant patients the mother would be the major beneficiary while the foetus can be at major risk.

In the justification of radiotherapy in pregnant patients, the proximity of the tumour to the foetus is a major factor in the decision. If the foetus is receiving only scattered radiation from the chest, the main concerns will be related to potential stochastic effects (mainly childhood cancer/leukaemia) and, depending on the stage of pregnancy and proximity of the treatment field, perhaps a decrease in the child's IQ. For treatment of tumours in the pelvic region the foetus will be in the primary radiation beam or very close to it and effects will typically be severe (usually foetal death).

Reference [39] states that "there are no hard and fast rules" for a decision on radiotherapy of pregnant patients. Other alternative or combined techniques are also associated with risks to the foetus. Chemotherapy using some drugs, administered in the third trimester of pregnancy, causes malformation of the foetus in 10% of cases and there are some suggestions that in-uterus exposure to chemotherapy may pose a risk of pancytopaenia at birth and possibly subsequent neoplasm in the offspring. The risks of surgery and anaesthesia during pregnancy are well known, the major problems being associated with hypotension, hypoxia and infection. The decision, therefore, on the treatment course should be "an informed one made by the patient, the husband, or the appropriate person(s), the treating oncologist, and other team members (e.g. surgeons, obstetricians, pharmacologists and others such as psychologists)".

With respect to medical research, the BSS (para. II.8) require that:

"The exposure of humans for medical research is deemed to be not justified unless it is:

(a) in accordance with the provisions of the Helsinki Declaration[16] and follows the guidelines for its application prepared by Council for International Organizations of Medical Sciences (CIOMS)[17] and WHO[18]; and
(b) subject to the advice of an Ethical Review Committee (or any other institutional body assigned similar functions by national authorities) and to applicable national and local regulations."

"[16] Adopted by the 18th World Medical Assembly, Helsinki, 1974, and as amended by the 29th World Medical Assembly, Tokyo, 1975, the 35th World Medical Assembly, Venice, 1983, and the 41st World Medical Assembly, Hong Kong, 1989; available from the World Medical Association, F-01210 Ferney-Voltaire, France."

"[17] COUNCIL FOR INTERNATIONAL ORGANIZATIONS OF MEDICAL SCIENCES in collaboration with WORLD HEALTH ORGANIZATION, International Ethical Guidelines for Biomedical Research Involving Human Subjects, CIOMS, Geneva (1993)."

"[18] WORLD HEALTH ORGANIZATION, Use of Ionizing Radiation and Radionuclides on Human Beings for Medical Research, Training and Non-Medical Purposes, Technical Report Series No. 611, WHO, Geneva (1977)."

Exposure to comforters shall be in accordance with the dose constraints formalized in the BSS (Schedule II-9). They shall be provided with instruction on actions to take to limit their exposure while visiting or caring for a patient who has received a brachytherapy implant.

5.3. OPTIMIZATION

According to BSS para. II.18, licensees of radiotherapy practices shall ensure that:

"(a) exposure of normal tissue during radiotherapy be kept as low as reasonably achievable consistent with delivering the required dose to the planning target volume, and organ shielding be used when feasible and appropriate;

(b) radiotherapeutic procedures causing exposure of the abdomen or pelvis of women who are pregnant or likely to be pregnant be avoided unless there are strong clinical indications;

....

(d) any therapeutic procedure for pregnant women be planned to deliver the minimum dose to any embryo or foetus;

(e) the patient be informed of possible risks."

Optimization of protection in pregnancy deserves special consideration. The International Commission on Radiological Protection (ICRP), in Publication 84 on Pregnancy and Medical Radiation [39], advises that "Termination of pregnancy is an individual decision affected by many factors. Foetal doses below 100 mGy should not be considered a reason for terminating a pregnancy. At foetal doses above this level there can be foetal damage, the magnitude and type of which is a function of dose and stage of pregnancy."

Calculation of the dose to the foetus before treatment of a pregnant patient should be part of the treatment plan. The distance from the field edge to the foetus is the most important factor in foetal dose, together with other factors such as field size, angle and radiation energy. The foetal dose from a typical photon treatment regimen for brain cancer can be in the range of 30 mGy. For anterior and posterior mantle treatments of the chest for Hodgkin's disease, the dose to portions of an unshielded foetus can be 400–500 mGy. Additional shielding can reduce the foetal dose, but effective shielding often weighs in the order of 200 kg, which must be considered in the design of the couch. The American Association of Physicists in Medicine has made a series of recommendations [40] for the planning, preparation and execution of the treatment.

Occasionally patients who are not pregnant ask when they can become pregnant after radiotherapy. Most radiation oncologists request that their patients not become pregnant one to two years after completion of therapy. This is not primarily related to concerns about potential radiation effects from the completed treatment, but rather to considerations about the risk of relapse that would require more radiation, surgery or chemotherapy.

The licensee needs to provide written instructions on actions to be taken to reduce exposure to comforters, caregivers and members of the public from sources in brachytherapy patients with permanent implants. These instructions should include minimizing prolonged contact with children and potentially pregnant women, and procedures to follow in the event that a source becomes dislodged.

5.3.1. Calibration

The BSS, in para. II.19, require that:

"Registrants and licensees shall ensure that:

(a) the calibration of sources used for medical exposure be traceable to a Standards dosimetry laboratory;

(b) radiotherapy equipment be calibrated in terms of radiation quality or energy and either absorbed dose or absorbed dose rate at a predefined distance under specified conditions, e.g. following the recommendations given in IAEA Technical Reports Series No. 277[20];

(c) sealed sources used for brachytherapy be calibrated in terms of activity, reference air kerma rate in air or absorbed dose rate in a specified medium, at a specified distance, for a specified reference date;

....

(e) the calibrations be carried out at the time of commissioning a unit, after any maintenance procedure that may have an effect on the dosimetry and at intervals approved by the [regulatory body].

"[20] INTERNATIONAL ATOMIC ENERGY AGENCY, Absorbed Dose Determination for Photon and Electron Beams, Technical Reports Series No. 277, IAEA, Vienna (1987)."

At the time of publication of the BSS, the IAEA code of practice based on air kerma free in air was included in Ref. [41], which provides the code of practice to determine the absorbed dose to water in the clinical beams using the ionization chamber calibrated in air kerma free in air at a Standards dosimetry laboratory. More recent codes of practice based on standards of absorbed dose to water, such as Ref. [42], were not available at that time. It is, however, necessary to extend the application of the BSS requirement to the new code of practice, which refers to the determination of absorbed dose to water in the hospital, using an ionization chamber also calibrated in absorbed dose to water at the secondary standards dosimetry laboratory.

Sealed sources used for external beam and brachytherapy need to have a calibration certificate provided by the manufacturer, in accordance with Ref. [43] or its national equivalent standards. However, the calibration of the beam and of the brachytherapy sources should be carried out by a qualified expert in radiotherapy physics before the source can be used to treat patients.

The licensee needs to implement a protocol for calibration of radiation sources used for radiotherapy. The regulatory body should encourage the professional medical physics bodies to adopt a protocol and require its implementation by licensees. It is advisable to use international protocols for calibration. This would avoid confusion and help prevent mistakes. Examples are the calibration procedures described by the IAEA in Refs [41, 42, 44] for external beam therapy and IAEA-TECDOC-1274 for brachytherapy [32]).

Ideally, two different qualified experts in radiotherapy physics, preferably using different dosimetry systems, should calibrate new equipment and new radiation sources independently and redundantly. The results should be compared only after completion of both measurements.[21]

The licensee should ensure that all teletherapy equipment outputs are compared at least once every two years in a national, regional or international programme for independent dose verification. One of the simplest mechanisms for independent verifications of external beam calibration or physical dosimetry is participation in the IAEA/WHO thermoluminescent dosimetry postal dose quality audit. The regulatory body should encourage registrants and licensees to participate in this or similar programmes.

New brachytherapy sources should be calibrated and variations in the measurement of more than 5% from the manufacturer's certified activity or kerma rate should be investigated. The sources should not be used for patient treatment until differences greater than 10% have been investigated and resolved. The responsibility for the investigation and for further action remains

---

[21] This verification of the calibration is essential but may not be straightforward in hospitals with only one medical physicist. In some countries there may be only one medical physicist. Experience shows, however, that it is a worthwhile measure and it may be advisable to arrange international assistance to ensure independent and redundant calibration when a new piece of equipment is commissioned or a ^{60}Co source is exchanged. Alternatively, the beam calibration should be checked through a postal TLD service, such as the IAEA/WHO postal service.

with the licensee, and the investigation is usually performed by the qualified expert in radiotherapy physics, with or without external help.

5.3.2. Clinical dosimetry

According to para. II.20 of the BSS,

"Registrants and licensees shall ensure that the following items be determined and documented:

....

(b) for each patient treated with external beam radiotherapy equipment, the maximum and minimum absorbed doses to the planning target volume together with the absorbed dose to a relevant point such as the centre of the planning target volume, plus the dose to other relevant points selected by the medical practitioner [the radiation oncologist] prescribing the treatment;
(c) in brachytherapeutic treatments performed with sealed sources, the absorbed doses at selected relevant points in each patient; [and]

....

(e) in all radiotherapeutic treatments, the absorbed doses to relevant organs."

To meet this requirement, i.e. the items to be determined and the way they are determined and documented, a protocol should be used. Similarly to the way in which protocols were recommended for calibration in Section 5.3.1, regulatory authorities, together with health authorities, should encourage professional bodies where available to develop or adopt a protocol. This should ensure complete and consistent treatment planning and recording. Two ICRU reports are recommended for this [45, 46].

5.3.3. Quality assurance for medical exposures

The BSS, in para. II.22, state that "Registrants and licensees, in addition to applying the relevant requirements for quality assurance specified elsewhere in the Standards, shall establish a comprehensive quality assurance programme for medical exposures with the participation of appropriate qualified experts in the relevant fields, such as radiophysics..., taking into account the principles established by the WHO[21–23] and the PAHO[24].

"[21] WORLD HEALTH ORGANIZATION, Quality Assurance in Diagnostic Radiology, WHO, Geneva (1982).

"[22] WORLD HEALTH ORGANIZATION, Quality Assurance in Nuclear Medicine, WHO, Geneva (1982).

"[23] WORLD HEALTH ORGANIZATION, Quality Assurance in Radiotherapy, WHO, Geneva (1988).

"[24] PAN AMERICAN HEALTH ORGANIZATION, Publicación Científica No. 499, Control de Calidad en Radioterapia: Aspectos Clínicos y Físicos, PAHO, Washington, DC (1986)."

The regulatory body should encourage licensees to work with professional associations in the development of such programmes. The licensee should ensure that the programmes are updated on a regular basis. As the development of a national programme may not be feasible in many Member States, a well established and proven international or national programme for QA may be followed [47, 48].

According to BSS para. II.23, "Quality assurance programmes for medical exposures shall include:

(a) measurements of the physical parameters of the radiation generators, imaging devices and irradiation installations at the time of commissioning and periodically thereafter;
(b) verification of the appropriate physical and clinical factors used in patient diagnosis or treatment;
(c) written records of relevant procedures and results;
(d) verification of the appropriate calibration and conditions of operation of dosimetry and monitoring equipment; and
(e) as far as possible, regular and independent quality audit reviews of the quality assurance programme for radiotherapy procedures."

Following the acceptance of new radiotherapy equipment, sufficient data should be collected at the commissioning to be used in treatment planning for clinical dosimetry. These data should be clearly documented in the workbook. Before being issued for use in treatment planning, the documentation should be independently verified, signed and dated. All dosimetry calibrations, clinical dosimetry data and methods of calculation for therapy equipment are to be reconfirmed at regular intervals. The measurements and checks carried out for this purpose should be sufficient to detect any significant variations from the data in use.

Verification of the treatment through in vivo dosimetry is advisable. This procedure may not be available in all institutions, but nevertheless it is recommended for incorporation as soon as it is feasible.

The QA programme should include auditing, both internal and external, and continuous improvement. The components of a comprehensive audit in radiotherapy are currently under scrutiny by the EU and the IAEA. These principles need to be linked to the radiation protection programme in order to strengthen safety while at the same time improving quality and efficiency. Feedback from operational experience and lessons learned from accidents or near misses can help identify potential problems and correct deficiencies, and therefore should be used systematically as part of the QA programme.

The maintenance of records is an important part of QA. When planning and developing an effective QA programme, licensees need to recognize that it demands strong managerial commitment and support in the form of training and time, personnel and equipment resources.

5.4. CONSTRAINTS FOR COMFORTERS AND VISITORS

Dose constraints do not apply to patients but do apply to the patients' comforters and visitors, whose exposure is also considered as medical exposure. Paragraph II.27 of the BSS establishes that: "Registrants and licensees shall constrain any dose to individuals incurred knowingly while voluntarily helping (other than in their occupation) in the care, support or comfort of patients undergoing medical diagnosis or treatment, and to visitors to patients who have received therapeutic amounts of radionuclides or who are being treated with brachytherapy sources, to a level not exceeding that specified in Schedule II, para. II-9."

Schedule II, para. II-9 quantifies this requirement by establishing that "the dose of any such comforter or visitor of patients shall be constrained so that it is unlikely that his or her dose will exceed 5 mSv during the period of a patient's diagnostic examination or treatment. The dose to children visiting patients who have ingested radioactive materials should be similarly constrained to less than 1 mSv." It is relatively straightforward to estimate effective doses to comforters and visitors from measurements of the ambient dose equivalent rates at the places where visitors are staying.[22]

---

[22] Ambient dose equivalent H*(10) can be used to estimate the personal dose equivalent Hp(10) that would correspond to an individual staying in the same field. Conversion coefficients are given in Ref. [36]. Hp(10) provides an estimate of the effective dose that avoids both underestimation and excessive overestimation.

5.5. DISCHARGE OF PATIENTS

The BSS, in para. II.28, establish that "In order to restrict the exposure of any members of the household of a patient who has undergone a therapeutic procedure with sealed...radionuclides and members of the public, such a patient shall not be discharged from hospital before the activity of radioactive substances in the body falls below the level specified in Schedule III, Table III-VI. Written instructions to the patient concerning contact with other persons and relevant precautions for radiation protection shall be provided as necessary." Table III-VI only refers to the value for ^{131}I (relevant only to therapeutic nuclear medicine) and sets 1100 MBq as the guidance level of maximum activity for patients in therapy on discharge from hospital.

An acceptable method to estimate the acceptable activity of permanent implants for patients being discharged from hospitals is to calculate the time integral of the ambient dose equivalent rate, considering the physical half-life of the radionuclides, and to compare this with the constraints on the patient's comforters, as discussed in Section 5.4, or for other persons who spend time in close proximity to the patient. Assumptions made in these calculations with regard to time and distance should be consistent with the instructions given to patients and comforters at the time of discharge of the patient from hospital. Examples of such calculations are given in Ref. [49]. The ICRP has an ongoing task group devoted to developing guidance for other sources, including those used for permanent implants in brachytherapy.

5.6. INVESTIGATION OF ACCIDENTAL MEDICAL EXPOSURE

According to the BSS:

"II.29. Registrants and licensees shall promptly investigate any of the following incidents:

(a) any therapeutic treatment delivered to either the wrong patient or the wrong tissue, or using the wrong pharmaceutical, or with a dose or dose fractionation differing substantially from the values prescribed by the medical practitioner [the radiation oncologist] or which may lead to undue acute secondary effects;

....

(c) any equipment failure, accident, error, mishap or other unusual occurrence with the potential for causing a patient exposure significantly different from that intended.

“II.30. Registrants and licensees shall, with respect to any investigation required under para. II.29:

(a) calculate or estimate the doses received and their distribution within the patient;
(b) indicate the corrective measures required to prevent recurrence of such an incident;
(c) implement all the corrective measures that are under their own responsibility;
(d) submit to the [regulatory body], as soon as possible after the investigation or as otherwise specified by the [regulatory body], a written report which states the cause of the incident and includes the information specified in (a) to (c), as relevant, and any other information required by the [regulatory body]; and
(e) inform the patient and his or her doctor about the incident.”

IAEA Safety Reports Series No. 17 [27] and ICRP Publication 86 [28] contain reviews of case histories from an extensive collection of accidental medical exposures. Appendix IV of this publication provides, for illustration, a short description of the major events and an exercise in identifying corrective measures for each case.

6. PUBLIC EXPOSURE

6.1. RESPONSIBILITIES

The licensee is responsible for controlling public exposure resulting from a radiotherapy practice. Public exposure is controlled by proper shielding design and, in large part, by ensuring that radiation sources are shielded and secured (e.g. located in a locked area), and that keys to the control panel are secured to prevent unauthorized access or use. The presence of members of the public in or near the radiotherapy department should be considered in the design.

The licensee should:

(a) Develop and implement use, storage and transport measures for ensuring the safety and security of radiotherapy sources to control public exposures in accordance with the requirements of the regulatory body;

(b) Control and maintain constant surveillance of licensed material that is not in storage (e.g. when brachytherapy sources are being transported or used for treatment) and secure stored licensed material from unauthorized access, removal or use (e.g. the storage facility should be locked at all times).

6.2. ACCESS CONTROL FOR VISITORS

The licensee should make arrangements to control access of members of the public to radiotherapy irradiation rooms and provide adequate information and instruction to these persons before they enter a controlled area, so as to ensure appropriate protection (e.g. members of the public should be accompanied).[23]

6.3. RADIOACTIVE WASTE AND SOURCES NO LONGER IN USE

The licensee should notify the regulatory body and submit a plan for transfer or disposal of sources if they are no longer in use. The licensee retains responsibility for the sources until the time of their transfer to another appropriate licensee or to an authorized waste disposal facility. Specifically, the licensee has to:

(a) Notify the regulatory body of any intention to transfer or decommission ^{60}Co teletherapy equipment prior to initiating an action. Depleted uranium used as shielding material should also be treated as radioactive waste. For example, a ^{60}Co teletherapy head may contain depleted uranium and is to be disposed of appropriately.
(b) Ensure that resources for the disposal of the sources will be made available when the teletherapy equipment is to be decommissioned.

Regulatory bodies may need to require applicants for licences to have in place a programme for safe disposal or return of the radioactive sources when their use is discontinued, before authorization for the import or purchase of equipment or radiation sources is given. A contract with the manufacturer or representative for the return of sources is acceptable evidence of such a programme.

---

[23] Exposure of comforters of patients is not public exposure, rather medical exposure. Nonetheless, comforters also need to be accompanied when entering controlled areas.

6.4. MONITORING OF PUBLIC EXPOSURE (BSS para. III.13)

"III.13. Registrants and licensees shall, if appropriate:

(a) establish and carry out a monitoring programme sufficient to ensure that the requirements of the Standards regarding public exposure to sources of external irradiation be satisfied and to assess such exposure;

....

(c) keep appropriate records of the results of the monitoring programmes."

The programme for monitoring public exposure from radiotherapy should include dose assessment in the areas of irradiation rooms for external beam therapy, of brachytherapy wards, and of source storage and preparation rooms and waiting rooms.

7. POTENTIAL EXPOSURE, MITIGATION AND EMERGENCY PLANS

The requirements for the safety of sources and facilities are set out in Section 3. This section focuses on identification of possible emergency or accident situations, their prevention, preparation for them and mitigation.

7.1. POTENTIAL EXPOSURE

7.1.1. Safety assessment

The BSS, in para. IV.3, require the licensee to conduct either a generic or a specific safety assessment of the sources for which they are responsible. The assessment is to be provided to the regulatory body, according to the BSS principal requirements on authorization (BSS paras 2.11–2.13). Basically, the safety assessment deals with finding 'what can go wrong' and how it can be prevented and, in case it occurs, how it can be mitigated.

Generic safety assessments are suitable for types of equipment with a high degree of uniformity. As an individual licensee's experience in identifying accident scenarios may be limited, arrangements between licensees and manufacturers to provide for notification of malfunctions and dissemination of feedback to licensees (see BSS para. IV.9) should be encouraged. In addition, dissemination of lessons from reported incidents and accidents should be encouraged by the regulatory authorities.

These measures can be complemented by participation in international mechanisms to share information. Finally, in order to ensure that the safety assessment is comprehensive and is not restricted to past events but also anticipates other possible events, consideration should be given to using systematic techniques, e.g. fault and event trees and probabilistic safety assessment techniques.

The assessment should be systematic and contain information on identification of possible events leading to accidental exposure (see Appendices C, D and E for a list of events, causes and contributing factors identified from real accidents). The safety assessment should not only cover these events, but also aim at anticipating other events that have not previously been reported.

The safety assessment should be documented and revised by an independent expert when:

(a) The radiation sources or their facilities are modified;
(b) Operational experience or information on accidents or errors indicates that the safety assessment should be reviewed;
(c) Techniques are modified in such a way that safety may be compromised.

7.2. ACCIDENT PREVENTION

The licensee has to incorporate:

(a) Defence in depth measures to cope with identified events, and evaluation of the reliability of the safety systems (including administrative and operational procedures, equipment and facility design).
(b) Operational experience and lessons learned from accidents [27, 28] and errors in training, maintenance and QA programmes. The licensee should promptly inform the regulatory body of all reportable events.

7.3. MITIGATION: EMERGENCY PLANS

In addition to the requirements in Appendix V of in the BSS, further requirements for emergency preparedness and response are given in Ref. [50]. Reference [51] provides guidance on the response to a range of radiological emergencies. The applicable parts of these documents were considered in the development of this section.

Based on the events identified by the safety assessment described in Section 7.1.1, the licensee should elaborate mitigation measures embodied in a set of emergency procedures, allocate responsibilities and provide for the training of the relevant staff in executing the mitigation measures, which should be rehearsed periodically. The lessons learned from the rehearsals should be used to review and update the emergency plans. The procedures should be concise and unambiguous, and should be posted visibly in places where they might be needed.

Only trained and authorized maintenance staff should handle incidents during source change of external beam therapy and remote control brachytherapy units. If the participation of radiotherapy staff is necessary for any of these actions, the scope of this participation should be limited to operating the equipment. The responsibilities of radiotherapy staff and maintenance staff for these specific situations should also be clearly defined.

Emergency plans should be developed by the licensee and approved by the regulatory body before the startup of a radiation treatment programme. The most frequent types of emergency situation are described in the following sections.

7.3.1. Lost radiotherapy source

It is critical for this type of event that an up to date inventory exists so that it can be determined immediately which sources are missing, their type and activity, when and where they where last known to be, and who last took possession of them. The area where the sources were last known to be should be closed to entry and exit until after a survey. This search needs to be performed with the most sensitive radiation detection survey meter available.

For sources over which control has been lost, the concept of a dangerous source is defined (see footnote 10). The following steps are recommended:

(a) Obtain assistance from the radiation protection officer;
(b) Conduct a local search;
(c) Check and ensure physical security and control of other sources;
(d) Report the theft or loss to the appropriate officials, providing a description of the device and its threat;
(e) Secure all information and the scene as much as possible to allow for forensic investigation;
(f) Conduct response actions in cooperation with local officials and law enforcement authorities;

(g) Identify and investigate routes by which the source may have been lost (e.g. waste, patient transfers);
(h) Brief off-site officials on risks and provide measures to protect emergency workers (including law enforcement personnel) and control their dose;
(i) Recommend that local officials inform nearby medical facilities, border crossings and scrap metal dealers to be alert for the source or for radiation induced injuries; provide them with a description of the source and its container and of symptoms of radiation injuries (e.g. burns with no apparent cause);
(j) Support local officials in explaining the risk to the local public and the media;
(k) Have the national regulatory body notify potentially affected States and the IAEA if there are indications that the source may have crossed into another State;
(l) If the source is found, ensure it is not damaged or leaking — if it is damaged or leaking, notify officials and ensure that it is surveyed for contamination;
(m) Reconstruct/record the doses received and inform those exposed of the risks; arrange, where appropriate, for long term medical follow-up.

7.3.2. Stuck sources

Emergency procedures need to be short, concise, unambiguous and, if necessary, illustrated with drawings without explanatory text. They need to be read at 'first sight' and followed. It has to be made clear that the first sight procedures refer to actions to be taken immediately to prevent/limit serious overexposures, or take other life saving actions [51]. Dose limits for occupational protection are not applicable to these actions. Further actions to recover the sources, to repair and test the equipment for returning it to use, are not directly part of the emergency response but occupational protection requirements, including dose limits, do apply to these actions.

In radiotherapy, however, the patient is directly in the radiation beam or brachytherapy sources are placed inside the patient; for this reason some of the emergency response actions coincide with source recovery actions, for example the retrieval of remote control brachytherapy sources from the patient to the safe, either electrically or using the manual crank.

7.3.2.1. External beam therapy units

Emergency procedures should be posted at the treatment unit. In the case of an event, the first step is generally to use the source driving mechanism to return the source to the shielded position. If this is not immediately successful and there is a patient on the treatment couch, the patient should be removed from the area and the area must be secured from further entry. Emphasis should be placed on avoiding exposure of the staff to the primary beam. The radiation protection officer is then notified and takes control of the situation.

Actions are to be performed only by staff that has been trained in the emergency actions, understands them and has regularly rehearsed. For external beam therapy the actions required have to be performed within seconds. Therefore the staff present in all procedures, i.e. the operators of the unit, have to be trained in these actions.

After the emergency actions, the following should be done:

(a) The maintenance engineer should be contacted to perform an inspection of the machine;
(b) The medical physicist should assess the patient doses and clear the use of the machine after maintenance;
(c) The radiation protection officer should assess the dose to the staff in the emergency and recovery operation;
(d) A record should be kept;
(e) The regulatory body should be notified.

7.3.2.2. Remote control brachytherapy units

Emergency plans require having an emergency container available in the treatment room, as well as an emergency kit containing long handled forceps for manipulation of the source guide tubes, and applicators if the source fails to return to the safe. The emergency container should be placed close to the patient and should be sufficiently large to accept the entire applicator assembly containing the source which has been removed from a patient.

For HDR brachytherapy, Ref. [11] states that: "High dose rate brachytherapy is potentially a high risk technique and extreme accuracy and care is essential. The short response time required for emergency actions (minutes) imposes the need for the presence of both a physician and physicist trained in emergency procedures during all applications."

Manufacturers usually provide suggested emergency procedures if the source fails to return to the safe. They assume that the physical integrity of the applicator is maintained. These procedures are specific to the actual afterloading unit but generally involve the following sequence. Each step assumes that if the previous action fails to lead to recovery, the following action is required. Actions are to be performed only by staff that has been trained in the procedure, understands it and has regularly rehearsed it. In the case of

HDR brachytherapy, the actions required have to be performed within seconds and therefore all operators need to be thoroughly trained on what to do in an emergency situation. The generic sequence is:

(1) Observation at the console of an error message and emergency indicators (audible and visible alarms);
(2) Recovery from the console (e.g. pressing an emergency off button);
(3) Entry into the room with a portable radiation survey meter (opening the door activates the interlock that retracts the source);
(4) Monitoring the radiation levels in the room;
(5) Recovery from the afterloading unit (by pressing an emergency off button on the remote afterloading unit);
(6) Manual retraction of the source (using a hand crank);
(7) Patient survey and afterloading survey (confirming that the source is in the safe);
(8) Applicator removal and placement in the emergency container;
(9) Patient survey and emergency container survey (to confirm that the source is not in the patient and is in the emergency container);
(10) Removal of the patient from the vault (with subsequent survey monitoring).

After the emergency the following should be done:

(1) The maintenance engineer should be contacted to perform an inspection and, if necessary, repair the machine;
(2) The medical physicist should make an assessment of the patient doses and clear the use of the machine after maintenance;
(3) The radiation protection officer should make an assessment of the dose to the staff in the emergency and recovery operation;
(4) The assessments should be recorded;
(5) The regulatory body should be notified.

7.3.3. Contamination

There is a very low probability of contamination accidents in radiotherapy departments where ^{226}Ra as well as old powder-form ^{137}Cs sources have been replaced. In case of a contamination accident it is important that the area be closed to further entry and that all who were in the area remain to be surveyed and decontaminated if necessary. If there are windows or other ventilation systems, these should be closed or turned off. There should also be a statement of how to contact the responsible radiation safety individual in the event of an emergency.

7.3.4. Off-site accidents

Off-site accidents with major consequences have been caused by loss of security of teletherapy sources not in use. Some of them (in Mexico and Brazil) caused large scale contamination and others caused external irradiation only (in Thailand and Turkey). Off-site accidents require actions by national intervention organizations. Participation of radiotherapy licensees in national emergency plans may be required.

In some cases, radiological emergencies involving public exposures from uncontrolled (lost or stolen sources) were first detected when medical professionals recognized the possibility of radiation induced injuries. Consequently there should be a procedure which includes whom to notify in the event that radiation induced injures are suspected among the public.

7.3.5. Accidental exposure of radiotherapy patients

The BSS requirements on investigation of accidental medical exposure are dealt with in Section 5.6, which includes investigation of the event, recording and reporting to the regulatory body and corrective measures to be taken. Appendix D contains an exercise on identifying corrective actions from major events.

8. SAFETY IN THE TRANSPORT OF RADIOACTIVE MATERIALS

It is a common practice for suppliers to transport external beam sources and remote control brachytherapy sources under their own responsibility (under their own licence), until the transfer of ownership of the source and acceptance tests are complete. Sources for manual brachytherapy are usually delivered directly to the hospital. In other cases it is the licensee of a radiotherapy department who makes all transport arrangements. The term 'licensee' in this section refers to the person responsible for the transport of the sources.

The licensee has to comply with the requirements of the IAEA Regulations for the Safe Transport of Radioactive Material [33] and/or any existing equivalent national legislation for all activities involving transport of radioactive sources. In the case of radiotherapy, this requirement applies to external beam radioactive sources and to brachytherapy sources (as indicated, unsealed sources are not the subject of this regulatory guidance).

8.1. RECEIPT OF RADIOACTIVE MATERIALS

Prior to each shipment of radioactive material to be dispatched, the licensee has to make the necessary arrangements with the source supplier to receive the relevant information. This information has to include the following for each package or container:

(a) The nuclide, number and activity of sources;
(b) Description of the source construction and performance tests, including leak tests;
(c) Special form approval certificate (where appropriate);
(d) Description of the package;
(e) Approval certificate for type A or B packages, or statement of compliance with Ref. [33] for other packages;
(f) Details of any special arrangements required, including multilateral approvals where necessary;
(g) A copy of the transport documents (to be sent to the licensee by fax or email before dispatch if possible).

The licensee should not agree to the dispatch of the consignment by the supplier unless all the above items are satisfactory. The supplier and licensee should agree on the transport route and the responsibility for each stage of the journey.

The following arrangements need to be made as appropriate and necessary:

(a) Special handling equipment for external beam sources, e.g. cranes, forklift trucks, etc., during transfer from one mode of transport to another, or between vehicles;
(b) Checking of radiation dose rates from the package or container;
(c) Checking that the correct transport labels are attached to the package or container and replacing any that are damaged or illegible;
(d) Ensuring that the package or container is securely attached to the vehicle and that the vehicle is correctly labelled;
(e) Dealing with border formalities;

(f) Security of the consignment during transport, particularly during delays or overnight stops.

8.2. DISPATCH OF RADIOACTIVE MATERIALS

The licensee should return packages or containers to the source supplier after receipt of a consignment of radioactive material. In the case of external beam therapy or remote control brachytherapy, packages or containers usually contain the disused radioactive sources.

8.3. EMPTY PACKAGES

With regard to returning empty packages the licensee has to:

(a) Carry out dose rate and contamination monitoring of both the inside and outside of the package or container to ensure that there is no residual radioactive material present and it can therefore be treated as an empty package or container;
(b) Remove or cover all transport labels relating to the sources contained in the package or container when received;
(c) Examine the package or container to ensure that it is in good condition and then close it securely, referring to any procedures provided by the source supplier;
(d) Attach a label to the outside of the package or container stating "UN 2908 RADIOACTIVE MATERIAL EXCEPTED PACKAGE – EMPTY PACKAGING";
(e) Complete a transport document;
(f) Contact the source supplier and agree to the transport route and responsibility for each stage of the journey; inform the source supplier of the proposed date of dispatch.

8.4. RETURN OF DISUSED SOURCES

With regard to returning disused sources, the licensee should provide the following information to the consignee for each package or container:

(a) The nuclide, number and activity of sources;
(b) A description of the source construction, including leak tests;

(c) Special form approval certificate (where appropriate);
(d) A description of the packaging in which the source is to be transported;
(e) Approval certificate for a type B package or statement of compliance with standards for other packages (as appropriate);
(f) Details of any special arrangements required, including multilateral approvals where necessary;
(g) A copy of the transport documents (to be sent to the consignee by fax or email before dispatch if possible).

The licensee should not dispatch the consignment unless confirmation has been received from the consignee that they are prepared to accept it. The licensee and consignee should agree on the transport route (as needed) and responsibility for each stage of the journey. The licensee will normally be responsible from dispatch until the consignment reaches the consignee's premises. Other arrangements are satisfactory provided they are agreed in advance by both parties and are also acceptable to the regulatory authorities. In order to prepare the consignment for dispatch the licensee should:

(a) Load the sources into the package, verifying the details to be provided to the consignee, e.g. serial numbers and comparable information to be entered on the transport document;
(b) Close the package or container securely and then examine it to ensure that it is in good condition, referring to any procedures provided by the source supplier;
(c) Carry out contamination monitoring of the outside of the package or container to ensure that there is no residual radioactive material present and it is therefore suitable for transport;
(d) Carry out dose rate monitoring of the package or container and attach appropriate transport labels;
(e) Refrain from using the transport labels relating to the sources contained in the package or container when received;
(f) Complete a transport document.

Arrangements should also be made for the following where necessary:

(g) Specify the need for special handling equipment, e.g. cranes, forklift trucks, etc., during transfer from one mode of transport to another, or between vehicles;
(h) Ensure that the package is securely attached to the vehicle and that the vehicle is correctly labelled;
(i) Deal with border controls;

(j) Provide security of the consignment during transport, particularly during delays or overnight stops.

In this situation the roles of licensee and source supplier are effectively reversed compared to Section 8.1, but the requirements are essentially the same.

Appendix I

SUMMARY OF ITEMS FOR REVIEW OF RADIATION PROTECTION AND SAFETY IN RADIOTHERAPY

Following is a list of major items to assist in appraisals of radiation protection and safety in radiotherapy. The relative complexity of each facility should be taken into account when assessing compliance. The list is only intended to provide a systematic approach to an appraisal, to ensure consistency in these appraisals and to avoid missing major items. It should not be construed as replacing professional judgement and knowledge of how safety features fit into the operation of a radiotherapy practice and of how to avoid interfering with medical care. The list can be used as guidance for self-assessment by the licensee, by peers when performing an appraisal and by regulators when checking compliance with the BSS.

I.1. GENERAL INFORMATION ABOUT THE FACILITY AND ADMINISTRATIVE REQUIREMENTS

- Patient workload (number of new cancer patients treated by radiotherapy/year, number of radiotherapy courses/year, number of radiotherapy sessions or fractions/year, number of radiotherapy fields treated/year):
  - External beam;
  - Brachytherapy;
- Treatment machines (number and type);
- Number of brachytherapy sources (specify type and number);
- Availability of an authorization granted by the regulatory body to build the facility, import the source and operate the radiotherapy practice;
- Specific conditions in the authorization;
- Previous reviews and inspections performed;
- Safety concerns in previous appraisals.

I.2. SECURITY OF SOURCES

- Provisions to keep an inventory of all sources in the radiotherapy department;
- Responsibilities for keeping and updating the inventory clearly assigned;

– Logbook to keep track of all movements of the sources; responsibility for keeping it assigned to a specific person;
– Provisions for dealing with disused sources in a safe manner (return to manufacturer or disposal — describe;
– Mechanism for prompt reporting of any missing sources, both internal to management and to the regulatory body;
– Means to prevent unauthorized access to the sources.

I.3. RADIATION PROTECTION AND SAFETY PROGRAMME

– Radiation protection and safety programme in place and endorsed by the licensee (the legal person);
– Radiation protection committee or equivalent mechanism;
– Members of the committee (usually it should include the chief radiation oncologist, a qualified expert in radiotherapy physics, a radiotherapy technologist, the radiation protection officer, a person responsible for coordinating maintenance and an administrator (representing the hospital management) for decision making and provision of resources);
– Clear definition of responsibilities in the radiotherapy department;
– Understanding of these responsibilities by the responsible staff and acknowledgement by them;
– Provisions to ensure that only qualified staff assume the above responsibilities.

I.4. RULES AND PROCEDURES

Procedures for:

– Purchasing radiation sources and radiotherapy equipment (preparation of technical specifications before purchasing, who should be involved and who provides the internal clearance);
– Receipt, storage and disposal of radioactive sources;
– Use of radiotherapy equipment, including safety devices;
– Individual exposure monitoring (see occupational protection);
– Workplace monitoring (see occupational protection);
– Leak testing;
– Communication of safety critical issues;
– Maintenance and repair of radiotherapy equipment, including obligatory notification to the qualified expert in radiotherapy physics before

resuming use (for a decision whether beam measurements are necessary before resumption of treatments);
— Moving radiation sources and patients with sources within the hospital.

I.5. PROTECTION AGAINST OCCUPATIONAL EXPOSURE

— Provisions to inform the workers about their obligations and responsibilities for their own protection and the protection of others against radiation and for the safety of sources.

Conditions of service

— Provisions to encourage pregnant workers to give notice of their pregnancy and to adapt their working conditions so as to ensure that the embryo or foetus is protected and afforded the same broad level of protection as required for members of the public, without excluding the female worker from work.

Classification of areas

— Controlled areas: all irradiation rooms for external beam therapy and remote afterloading brachytherapy, operating rooms during brachytherapy procedures using real sources, brachytherapy patient rooms, radioactive source storage and handling areas.

Local rules and supervision

— Procedures for ensuring adequate levels of protection and safety of workers;
— Provisions to make sure that these procedures, the protective measures and safety provisions are known to those workers to whom they apply and to other persons who may be affected by them;
— Supervision to ensure observance of the procedures;
— Investigation levels in place;
— Cooperation between workers and employers and licensees of both facilities, in cases where some workers are employed in other facilities using radiation — describe.

Personal protective equipment

- Availability of tools and devices for protection of workers in place (interlocks, tools for handling brachytherapy sources, mobile shielding, etc.) — describe.

Individual monitoring, exposure assessment and workplace monitoring

- Arrangements to have individual monitoring provided by an accredited and authorized service;
- Identification of staff members requiring individual monitoring;
- Establishment of the monitoring period, frequency of readings, recording of the accumulated doses and rules for returning and changing dosimeters;
- Arrangements to ensure that information on doses is made available to the staff;
- Rules for estimating the worker's dose if a personal dosimeter is lost or damaged.

Monitoring the workplace

- Provisions for keeping the workplace under supervision and the monitoring at a frequency that enables assessment in controlled areas and supervised areas.

Health surveillance

- Arrangements for health surveillance based on the general principles of occupational health;
- Counselling for pregnant women available.

Records

- Provisions for keeping records of each worker for whom assessment of occupational exposure is required.

I.6. PROTECTION AGAINST MEDICAL EXPOSURE

Responsibilities and training

- Assignment of the overall responsibility for patient protection and safety to a medical practitioner — specify (department head, radiation oncologist, etc.);
- Assignment of the responsibility for conducting or supervising calibration of beam and sources, clinical dosimetry and QA to a qualified expert in radiotherapy physics — specify type of expert (medical physicist specialized in radiotherapy, hospital physicist);
- Provision to review this number when workload increases, when new equipment is purchased or new techniques are introduced — specify;
- Documented education and training of all staff;
- Provisions for additional training when needed (new equipment, new techniques).

Justification of medical exposure

- Procedure to ensure and provide evidence that a decision to apply a therapeutic medical exposure is made by a radiation oncologist;
- Provisions for a formal justification before performing research that involves application of radiation on humans, according to the Declaration of Helsinki.

Optimization: Design and testing

- Acceptance test carried out according to international (such as IEC) or equivalent national standards for radiotherapy equipment — describe;
- Programme of commissioning in place, including treatment equipment as well as TPSs and simulators and other ancillary equipment — describe.

Optimization: Operational considerations

- Provision for optimization (exposure of normal tissue during radiotherapy to be kept as low as reasonably achievable consistent with delivering the required dose to the planning target volume, and organ shielding to be used when feasible and appropriate), for example:
  - Fixation devices used to reproduce treatments;
  - Checks that the position of the patient at the radiotherapy unit agrees with that for the dose planning;

- Portal films taken to verify the treatment;
- Participation of the radiation oncologist and qualified expert in radiotherapy physics in the first patient set-up;
- Other.

Optimization: Calibration

– Provisions for calibration of radiation beams and brachytherapy sources.
– Redundant independent verification as part of the provisions.
– Internationally accepted protocol or code of practice for calibration (absorbed dose determination to reference point) in place? Which one?
– Programme for follow-up calibration in place — describe.
– Participation in a dose quality audit programme.
– Provisions for source activity verification and identification of brachytherapy sources before use.
– Calibration programme: at the time of commissioning a unit, after any maintenance procedure that may have an effect on the dosimetry and at intervals approved by the regulatory body.

Optimization: Clinical dosimetry

– Procedure in place for specifying the absorbed doses to the target and to relevant organs — describe;
– Provisions for cross-checks of dose calculations?

Optimization: QA

– QA programme based on widely accepted and proven protocols — describe;
– Assignment of all tasks of the programme assigned to qualified persons;
– Availability of necessary instruments, quality control equipment and other ancillary equipment, as described in the programme;[24]
– Considerations or provisions for external audits as part of the programme;
– Maintenance programme, including follow-up of any safety related equipment fault detected by quality control or by any other means;
– Provisions to ensure that brachytherapy sources do not remain in the patient, including monitoring of patients and clothing.

[24] It is advisable to consider the feasibility of implementing in vivo dosimetry.

Investigation of accidental medical exposure

- Provision in place to investigate and report:
  - Any treatment delivered to the wrong patient, the wrong tissue, or with a dose or dose fractionation differing substantially from the values prescribed by the radiation oncologist or which may lead to undue secondary effects;
  - Any equipment failure, accident, error, mishap or other unusual occurrence with the potential to cause a patient exposure significantly different from that intended;
- Provision to estimate the doses received, indicate corrective measures to prevent recurrence, implement the corrective measures, submit a report to the regulatory body and inform the patient.

I.7. PROTECTION AGAINST PUBLIC EXPOSURE

- Provisions for protection of the public under normal operating conditions through:
  - Shielding and control of access and visitors;
- Provision to reduce the likelihood of accidents involving the public through:
  - Warning signals;
  - Provisions to ensure that control of sources is never relinquished;
  - Ensuring safe transport;
  - Safely dealing with disused sources (see items under Security, above).

I.8. EMERGENCY PREPAREDNESS AND RESPONSE

- List of predictable incidents and accidents and measures to deal with them;
- Persons responsible for taking actions, with complete relevant information, including telephone numbers;
- Responsibilities of individuals in emergencies defined in procedures (radiation oncologist, medical physicists, radiation technologists, etc.);
- Set of concise instructions posted in a visible area;
- Availability or quick access to persons responsible for carrying out emergency response action;
- Equipment and tools necessary to carry out the procedures;
- Training and periodic rehearsal;

- Recording and reporting system;
- Immediate measures to avoid unnecessary radiation doses to patients, staff and the public (such as removal of patients from a teletherapy unit, removal of implants, return of sources to the shielded position in remote control brachytherapy and teletherapy);
- Measures to prevent access of persons to the affected area during the time that the sources are exposed and normal conditions are restored;
- In the case of leaking sources, measures to prevent dispersion of contamination and access of persons to the contaminated area.

I.9. TRANSPORT OF RADIOACTIVE SOURCES

- Provisions to ensure that transport outside the hospital (for example, for returning sources) follows the IAEA Transport Regulations [33].

Appendix II

DOSE LIMITS FOR OCCUPATIONAL AND PUBLIC EXPOSURE

The following is reproduced from the BSS:

"OCCUPATIONAL EXPOSURE

"Dose limits

"II-5. The occupational exposure of any worker shall be so controlled that the following limits be not exceeded:

(a) an effective dose of 20 mSv per year averaged over five consecutive years[38];
(b) an effective dose of 50 mSv in any single year;
(c) an equivalent dose to the lens of the eye of 150 mSv in a year; and
(d) an equivalent dose to the extremities (hands and feet) or the skin[39] of 500 mSv in a year.

"II-6. For apprentices of 16 to 18 years of age who are training for employment involving exposure to radiation and for students of age 16 to 18 who are required to use sources in the course of their studies, the occupational exposure shall be so controlled that the following limits be not exceeded:

(a) an effective dose of 6 mSv in a year;
(b) an equivalent dose to the lens of the eye of 50 mSv in a year; and
(c) an equivalent dose to the extremities or the skin[39] of 150 mSv in a year.

"Special circumstances

"II-7. When, in special circumstances[40], a temporary change in the dose limitation requirements is approved pursuant to Appendix I:

(a) the dose averaging period mentioned in para. II.5 (a) may exceptionally be up to 10 consecutive years as specified by the Regulatory Authority, and the effective dose for any worker shall not exceed 20 mSv per year averaged over this period and shall not exceed 50 mSv in any single year, and the circumstances shall be reviewed when the dose accumulated by any worker since the start of the extended averaging period reaches 100 mSv; or

(b) the temporary change in the dose limitation shall be as specified by the Regulatory Authority but shall not exceed 50 mSv in any year and the period of the temporary change shall not exceed 5 years.

"PUBLIC EXPOSURE

"Dose limits

"II-8. The estimated average doses to the relevant critical groups of members of the public that are attributable to practices shall not exceed the following limits:

(a) an effective dose of 1 mSv in a year;
(b) in special circumstances, an effective dose of up to 5 mSv in a single year provided that the average dose over five consecutive years does not exceed 1 mSv per year;
(c) an equivalent dose to the lens of the eye of 15 mSv in a year; and
(d) an equivalent dose to the skin of 50 mSv in a year.

"Dose limitation for comforters and visitors of patients

"II-9. The dose limits set out in this part shall not apply to comforters of patients, i.e. to individuals knowingly exposed while voluntarily helping (other than in their employment or occupation) in the care, support and comfort of patients undergoing medical diagnosis or treatment, or to visitors of such patients. However, the dose of any such comforter or visitor of patients shall be constrained so that it is unlikely that his or her dose will exceed 5 mSv during the period of a patient's diagnostic examination or treatment. The dose to children visiting patients who have ingested radioactive materials should be similarly constrained to less than 1 mSv.

"[38] The start of the averaging period shall be coincident with the first day of the relevant annual period after the date of entry into force of the Standards, with no retroactive averaging.

"[39] The equivalent dose limits for the skin apply to the average dose over 1 cm^2 of the most highly irradiated area of the skin. Skin dose also contributes to the effective dose, this contribution being the average dose to the entire skin multiplied by the tissue weighting factor for the skin.

"[40] See Appendix I: the provisions for 'alternative employment' set out in para. 4.18 may be relevant."

Appendix III

EMERGENCY PROCEDURES

An emergency plan should include:

- A list of predictable incidents and accidents and measures to deal with them;
- Persons responsible for taking action with complete relevant information, including telephone numbers;
- The responsibilities of individual personnel in emergency procedures (radiation onocologist, medical physicists, radiation technologists, etc.);
- A set of concise instructions posted in a visible area;
- Availability or quick access to persons responsible for carrying out emergency response actions;
- Equipment and tools necessary to carry out the procedures;
- Training and periodic rehearsal;
- A recording and reporting system.

Emergency procedures should also include:

- Immediate measures to avoid unnecessary radiation doses to patients, staff and the public (such as removal of patients from a teletherapy unit, removal of implants, return of sources to the shielded position in remote control brachytherapy and teletherapy);
- Measures to prevent access of persons to the affected area during the time that the sources are exposed and normal conditions are restored;
- In the case of leaking sources, measures to prevent dispersion of contamination and access of persons to the contaminated area.

Appendix IV

CASE HISTORIES OF MAJOR ACCIDENTS INVOLVING ABANDONED RADIOTHERAPY SOURCES AND EXPOSURES TO MEMBERS OF THE PUBLIC

IV.1. ILLEGAL IMPORT AND INSECURE STORAGE OF AN EXTERNAL BEAM Co UNIT (MEXICO, 1984)

A second hand teletherapy unit with a 1000 Ci (37 TBq) ^{60}Co source was purchased and imported, but did not comply with all the existing import regulatory requirements. It was stored in a warehouse for six years, when its scrap value attracted the attention of a maintenance technician who dismantled the head of the unit and removed the cylinder containing the ^{60}Co [5].

The technician removed the cylinder and other metal parts from the unit, loaded them into a pickup truck, drove to a scrapyard and sold the parts as scrap. Before arriving at the scrapyard, he deliberately ruptured the cylinder containing the source. The pickup truck thus contained a large quantity of radioactive material from the source. The source consisted of about 6000 tiny (1 mm in diameter) pellets 1 mm of ^{60}Co. When the cylinder was ruptured, several pellets were dispersed and remained in the truck when the heavy parts were unloaded at the scrapyard. Owing to mechanical defects, the pickup truck contaminated with ^{60}Co pellets remained parked on the street for 40 days. The truck was then moved to another street where it stood for a further ten days.

When the ruptured cylinder was moved by cranes in the scrapyard, together with the other metal pieces, the ^{60}Co pellets were spread over the scrapyard, attracted by the crane's magnetic field, and mixed with the other metal materials. Consequently, pellets and pellet fragments also were transferred to other vehicles used for transporting scrap to various foundries. The main purchaser of the scrap was a firm that manufactured construction reinforcing rods and connecting rods for motor vehicles.

It was later found that scrap contaminated with ^{60}Co had been used by steel production plants to manufacture reinforcing rods and metal table bases. A lorry transporting contaminated rods to be used for construction entered a road leading to a nuclear laboratory where radiation detectors were used to monitor the removal of radioactive material. The detectors not only indicated the presence of radioactivity but also activated a camera that photographed the contaminated vehicle. Two days later the authorities ascertained the origin of the contaminated rods.

An extensive investigation showed that 30 000 table bases and 6600 tonnes of reinforcing rods had been made from the contaminated material. Aerial radiation surveys of an area of 470 km^2 were conducted, resulting in the recovery of 27 cobalt pellets. Visits were made to 17 636 buildings to determine whether contaminated material had been used in their construction. Unacceptable radiation levels were measured in 814 buildings that were then partly or completely demolished. The accident exposed approximately 4000 people to radiation, about 80 of whom received doses higher than 250 mSv. Apparently five people received doses of 3–7 Sv over a two month period.

Although the triggering event was a person dismantling an insecurely stored head of a ^{60}Co teletherapy unit and breaking the source capsule, other factors contributed, such as non-compliance with regulations, as the radiotherapy unit had been illegally imported and transported and remained in storage for six years under unsafe and unsecured conditions.

IV.2. ABANDONMENT OF AN EXTERNAL BEAM ^{137}Cs UNIT (BRAZIL, 1988)

A private radiotherapy institute moved to new premises leaving a caesium (^{137}Cs) teletherapy unit behind without notifying the licensing body. Because of partial demolition of the building, the teletherapy unit became totally unsecured. Two people entered the building and removed the source assembly from the radiation head. They tried to dismantle the source assembly at home and in the attempt the source capsule was ruptured. Because the radioactive source was in the form of caesium-chloride powder, which is highly soluble and easily dispersed, there was considerable contamination of the environment, resulting in external irradiation and internal contamination of several persons.

After the source capsule was ruptured, remnants of the source assembly were sold for scrap to a junkyard owner. He noticed that the source material glowed blue in the dark. This fascinated several persons and over a period of days, friends and relatives came and saw the phenomenon. Fragments of the source the size of rice grains were distributed to several families. Five days later, a number of persons showed gastrointestinal symptoms that were not recognized initially as being due to exposure to radiation. However, one of the irradiated persons connected the illnesses with the source capsule and took the remnants to the public health department. This action started a chain of events that led to investigation of the accident.

Many individuals received external and internal radiation exposures exceeding acceptable limits. In total, some 112 000 persons were monitored; 249 persons were contaminated either internally or externally. Some suffered very high internal and external contamination owing to the way they had handled the caesium-chloride powder, such as daubing their skin, eating with contaminated hands, and handling various objects. Four persons died within four weeks of admission to hospital, having received total body doses of at least 5 Gy.

Seven main sites of contamination were identified, including the junkyards concerned, some of them with dose rates of up to 2 $Sv \cdot h^{-1}$ at 1 m. Aerial surveys were conducted over 67 km^2. Of 159 houses monitored, 42 required decontamination. The final volume of waste stored was 3500 m^3 or 275 lorry loads. The radioactivity accounted for in the decontamination was estimated at 1200 Ci (44.4 TBq), compared with the known activity of the source before the accident of 1375 Ci (51 TBq).

This case is similar to that discussed in Section 2.3. Lack of compliance with regulations contributed to the accident (there was no safe, formal and complete decommissioning of the facility, and the ^{137}Cs teletherapy unit which was no longer in use was abandoned unsecured).

FIG. 1. Boxes and drums of waste stacked and covered at a temporary storage site.

IV.3. TELETHERAPY SOURCES STORED IN A GENERAL PURPOSE WAREHOUSE AND LEFT BEHIND IN THEIR TRANSPORT CONTAINERS (TURKEY, 1998–1999)

A company based in Ankara was licensed by the regulatory body to import, transport and re-export radioactive sources. In 1993 the company is recorded as having loaded three disused radiotherapy sources into individual type B(U) transport packages in preparation for returning them to the original supplier in the USA. The regulatory body had given permission to transport and export the packages, but the company did not send the sources but stored them in Ankara from 1993 to 1998 without informing the regulatory body.

In February 1998, the company transported two of the packages from Ankara to Istanbul and stored them in their general purpose warehouse on an industrial estate. After some time there was no room in this warehouse and the packages were moved to adjoining premises that were empty. After nine months or so the adjoining premises were transferred to new ownership and the new owners, not realizing what was in the packages, sold both of them as scrap metal. The packages were labelled with the trefoils but the persons who purchased them were unaware of the radiation hazard. They opened them and

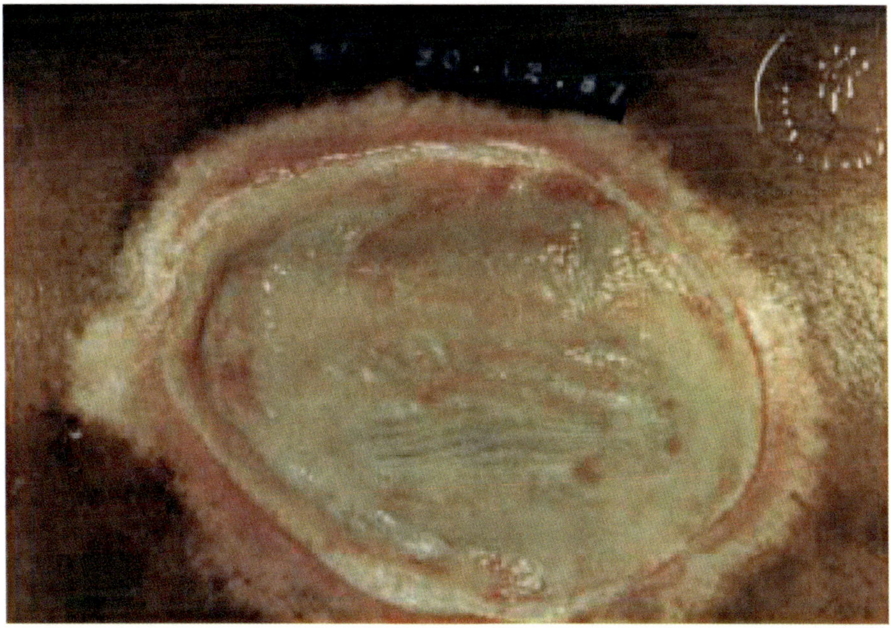

FIG. 2. Injury after extension into the musculature of the thigh.

broke open the shielded containers, thereby unwittingly exposing themselves and several others to radiation from at least one unshielded ^{60}Co source. This occurred in a residential area of Istanbul on 10 December 1998.

A total of ten persons who had spent time in the proximity of the dismantled containers fell ill and six of them began to vomit. Although they sought medical assistance, the cause of the illness was not recognized until almost four weeks later (on 8 January 1998). Over a period of about two weeks, pieces of the dismantled containers and at least one of unshielded source were left in a residential area before being taken to a local scrapyard, where they were left for a further two weeks. When the injuries were eventually suspected as having been caused by radiation exposure, the doctors immediately alerted the national authorities. As a result, one unshielded source was quickly discovered at the scrapyard and safely recovered, thus preventing further radiation exposure.

As a result of media reports, a total of 404 persons applied for medical checks because of fears about possible radiation exposure. Of these, a total of 18 persons (including seven children) were admitted to hospitals. Ten adults exhibited clinical symptoms of acute syndrome. The five most exposed were hospitalized for 45 days.

FIG. 3. The scrapyard at Ziya Gökalp.

IV.4. INSECURE STORAGE OF A DISUSED TELETHERAPY UNIT IN SAMUT PRAKARN (THAILAND, 2000)

The teletherapy device involved in the accident had originally been installed at a hospital in Bangkok in 1969 and the source was replaced in 1981. The hospital did not obtain any further service from the manufacturer, and the manufacturer's local agent who had been contracted to do maintenance on the unit later declared bankruptcy. The teletherapy unit was reported to have been taken out of service in 1994. The hospital purchased a new unit but the manufacturer of the new unit did not accept the return of the disused source.

The hospital was left with a disused source to manage and control as required by the regulatory body. Since the hospital did not have storage space it sold the device and source to the agent of the new supplier and did not inform the regulatory body of this transfer. This company had another teletherapy source which it had been authorized to store at a warehouse that they leased in the Bangkok area.

Early in 1999 the company was notified that its lease of the warehouse was to be terminated and it relocated this unit, together with other sources that it possessed, to a car parking lot that was owned by a parent company. The car parking lot was fenced but there were gaps in the fences and residents across the street played football in the open space of the parking lot near the storage place of the sources.

In January 2000, several individuals obtained access to the storage location and partially disassembled a teletherapy head. They took the unit to the residence of one of the individuals, where four people attempted to disassemble it further. The teletherapy head displayed a radiation trefoil and warning label. However, the individuals did not realize that this indicated radioactive material and the warning label was not in a language they understood. On 1 February 2000, two of the individuals took the partially disassembled device to a junkyard in Samut Prakarn so that the component metals could be segregated and sold separately. While a worker at the junkyard was disassembling the device using an oxyacetylene torch, the source fell out of its housing, unobserved by either the junkyard workers or the individuals involved.

By the middle of February 2000, several of the individuals involved had begun to feel ill and sought medical assistance. Physicians at a local hospital recognized the symptoms of several of the patients and suspected that an unsecured radiation source was the cause. They reported their suspicions to the regulatory body. Officials from the regulatory body, assisted by local public health personnel, searched for the source. When high radiation levels were found in the vicinity of the junkyard they secured the area to prevent further

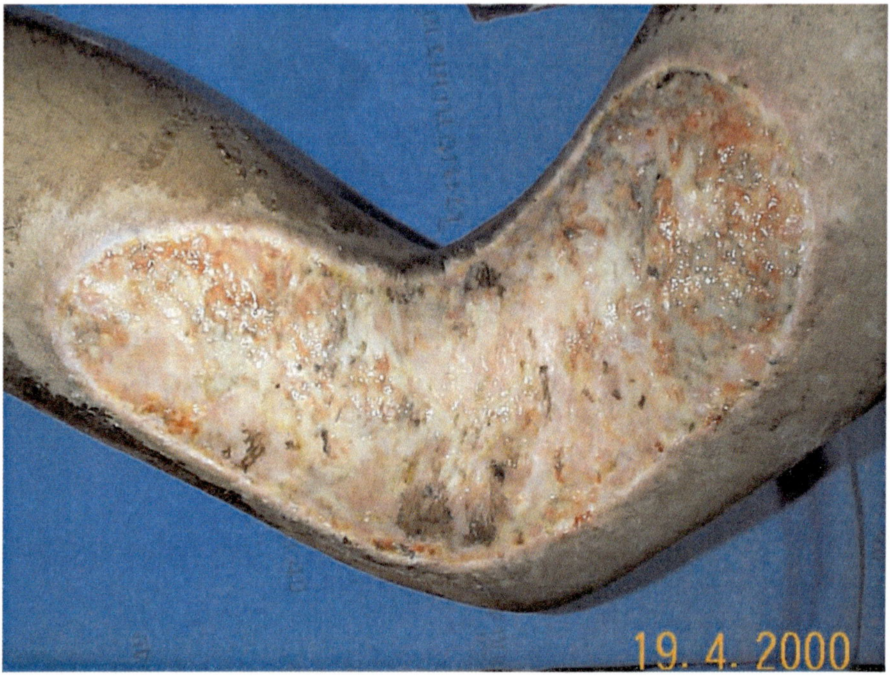

FIG. 4. Massive infection and beginning of necrosis of an extended wound on the right leg on 19 April 2000, 11 weeks after exposure.

access. An emergency response team was assembled and by 20 February 2000 the source was recovered and transported to a secure storage area.

Ten individuals were severely exposed and developed burns, nausea, vomiting and depilation. Four of these individuals received more than 6 Gy whole body doses and three of them died within two months of the accident as a consequence of their severe radiation injuries.

Appendix V

CASE HISTORIES OF SOME MAJOR ACCIDENTAL MEDICAL EXPOSURES IN RADIOTHERAPY

References [27, 28] contain information on about 90 events. The most severe of these are shown in Table 3 and some of the major ones are described in detail in the subsequent sections.

V.1. INCORRECT ^{60}Co DECAY CHART AND LACK OF VERIFICATION (USA, 1974–1976)

A cobalt (^{60}Co) teletherapy unit had initially been calibrated correctly, but subsequent dose calculations were based on an incorrect ^{60}Co decay curve which showed a more rapid decay than the real decay of the source. The dose rate was therefore underestimated and, consequently, treatment times derived from the erroneous decay curve were longer than necessary.

Initially the medical physicist tried to attribute the problem to a faulty measuring system and produced ten calibration documents showing that the report of machine output used was correct. When some patients showed symptoms of overexposure, the hospital asked external consulting physicists to review the dosimetry. They found that the ^{60}Co unit and measuring system were functioning correctly, but that the resident medical physicist had fabricated all but one of the machine output reports.

Overdose to patients had increased progressively over 22 months as the difference between the real decay and the incorrect curve increased. The level of overdose was 10% in the fifth month and increased to as high as 50% by the end of the 22 month period. During the last 16 months, 426 patients were treated. Approximately 20 years after the accidental exposure, Cohen et al. [52] evaluated the clinical consequences using the medical records of 450 patients. The clinical follow-up notes pertaining to the period between one and three years after the treatment showed that a high proportion of patients developed significant, often life threatening, complications. These included severe skin reactions with ulceration, mucosal reactions with necrosis, stenosis of pharynx and oesophagus, ulceration and/or perforation of stomach and bowel, bone necrosis and myelopathy.

A contributory factor may have been the insufficient staff resources to ensure adequate dosimetry and quality control. The actual beam output was not checked over a 22 month period and the medical physicist was assigned to the

TABLE 3. MAJOR REPORTED ACCIDENTAL EXPOSURES INVOLVING PATIENTS UNDERGOING EXTERNAL BEAM RADIOTHERAPY AND BRACHYTHERAPY TREATMENTS

Country	Year	Number of patients affected	Causes and main contributing factors
USA (see Section V.1)	1974–1976	426	^{60}Co dose calculations based on erroneous decay curve (varying overdoses) No independent verification of dose calculations More than two years without beam measurements
Canada and USA (see Section V.2)	1985–1987	Six accidental exposures, three deaths from radiation	Accelerator software for the control of treatment was transferred from other equipment without sufficient consideration of safety
Germany	1986–1987	86	^{60}Co dose calculations based on erroneous dose tables (varying overdoses) No independent determination of the dose rate
UK	1988	207	Error in the calibration of a ^{60}Co therapy unit (25% overdose) No independent calibration of the beam
UK	1988–1989	22	Error in the identification of ^{137}Cs brachytherapy sources (dosimetry errors between –20% and +10%) No independent determination of source strength
USA (see Section V.3)	1987–1988	33	A computer file (of a TPS) for the use of trimmers was not updated for a new ^{60}Co source because it was not intended for further use The computer file was used (overdoses of 75%) No manual verification of the calculated treatment time No independent verification of the treatment time

TABLE 3. MAJOR REPORTED ACCIDENTAL EXPOSURES INVOLVING PATIENTS UNDERGOING EXTERNAL BEAM RADIOTHERAPY AND BRACHYTHERAPY TREATMENTS (cont.)

Country	Year	Number of patients affected	Causes and main contributing factors
Spain (see Section V.4)	1990	27 (15 deaths directly from radiation; two deaths with radiation as a major contributor)	Error in the maintenance of a clinical linear accelerator Procedures for transferring machine from/to maintenance (notification of physicists) not followed Conflicting signals and displays not analysed Procedures for periodic beam verifications not implemented or insufficient Overdoses ranging from 200% to 600%
UK (see Section V.5)	1982–1991	1045 (492 patients developed a local recurrence, possibly as a result of the underdose)	Inappropriate commissioning of a computerized radiotherapy TPS (5–30% underdose) Lack of quality control
USA (see Section V.6)	1992	One (death from radiation)	HDR source left inside the patient and source dislodged from equipment Conflicting monitor signals (display at the machine and area monitor) ignored A survey with an available portable instrument was not conducted
Costa Rica (see Section V.7)	1996	114 (17 deaths from radiation)	Error in the calibration of a ^{60}Co therapy unit Lack of independent beam calibration and quality control Recommendations of external audit ignored Overdoses of about 60%
Panama (see Section V.8)	2000	28 (several deaths from to radiation)	Modified procedure for data entry into the treatment planning computer without verification prior to treatment
Poland (see Section V.9)	2001	Five patients with severe injuries	Failure of more than one layer of safety in an electron accelerator (failure of the power supply to the monitor chambers and of the relevant interlock)

support of a new accelerator. There was no independent check of the decay curves; the physicist subsequently falsified records in an attempt to justify his calculations.

Exercise to identify lessons and preventive actions from this event:

- What procedures does my radiotherapy department have for validating the basic data on which treatment calculations are based?
- If a mistake in the basic data were made in my department, would it be detected by an independent check?
- What provision do I have to re-evaluate staff resources and training for new equipment, new techniques, and when workload increases?

V.2. SIX ACCIDENTAL EXPOSURES INVOLVING SOFTWARE PROBLEMS IN SEVERAL ACCELERATORS OF THE SAME TYPE (CANADA AND USA, 1985–1987)

Between June 1985 and January 1987, six known accidental exposures involved massive overdoses from the same type of accelerator [53]. Several flaws were identified in the software used to enter selected parameters for the treatment, such as type of radiation and energy. Two of these flaws caused the deaths of three patients.

After a selection of the treatment parameters for the X ray mode, the operator changed the selection from X ray to electron mode. During this time the equipment was executing the initial request, which required the automatic rotation of a turntable assembly with the bending magnet for beam scanning (electron mode) and the target and beam flattener (X ray mode). The equipment did not execute the second request correctly and the turntable had not reached the correct position. The equipment then operated without the X ray target and the beam flattener, thus leading to massive overdoses.[25] An accidental exposure occurred in March 1986, the outcome of which was radiation induced myelitis of the cervical cord causing paralysis of the patient's left arm and both legs, left vocal cord paralysis (which left the patient unable to speak), neurogenic bowel and bladder, and paralysis of the left diaphragm. The patient died of complications from the overdose five months after the accidental exposure.

---

[25] The current in the linear accelerators is more than two orders of magnitude higher when operating in X ray mode than in electron mode to compensate for the loss of energy fluence in the X ray target and in the flattener.

The maintenance engineer could not reproduce the malfunction and the problem remained unresolved. The same problem reappeared in April 1986 in the same hospital and a second patient was overexposed. He developed disorientation that progressed to coma and neurological damage from high dose radiation injury to the right temporal lobe of the brain and the brain stem. The patient died three weeks after the accidental exposure.

The medical physicist in the hospital was finally able to reconstruct the fault, forcing the manufacturer to respond [54]. The exact doses in these cases are uncertain, but they were estimated to be of the order of 150–250 Gy per single treatment fraction (delivered within a few seconds).

A different software problem caused the death of a patient in another hospital in 1997. An operator pressed the 'set' button at the precise moment that a variable in the software rolled over zero. The machine started irradiation without an X ray target and without scanning of the electrons over the radiation field. The result was a highly concentrated electron beam.

There were a number of problems with the design, testing and process for following up equipment faults and in the dissemination of a warning to users of the same type of equipment. It is not possible to summarize these problems in a few lines. Interested readers are referred to detailed reports [53, 54]. Two of the major problems were:

(1) The software package for controlling some of the accelerator's functions was reused from an older type of accelerator with a significantly different design. In the new accelerator type, safety of these functions relied entirely on the old software.
(2) There was no efficient mechanism to follow up reports of suspected accidental exposures. The first accidental exposure occurred in June 1985 and resulted in a mastectomy due to complications from radiation. The patient completely lost the use of her shoulder and arm. In spite of this, the manufacturer and operators of the machine refused to believe that this could have been caused by the accelerator [53]. The accidental exposure was not reported to the US Food and Drug Administration until 1986.

Exercise to identify lessons and preventive actions from this event:

Manufacturers:

- What QA system do I have for design of radiotherapy equipment to ensure compliance with IEC or equivalent national standards?

– What system do I have for efficient reporting and follow-up to achieve complete clarification of unusual equipment performance and timely reporting to all users of equipment that might be affected?

V.3. COMPUTER FILE NOT UPDATED FOR ^{60}Co SOURCE CHANGE (USA, 1987–1988)

A computer program file used for the treatment of brain cancer with 'trimmer bars' was not updated in March 1987 when the therapy department replaced the ^{60}Co source. Other program files had been updated, but this one had not because trimmer bars were not used at that time to treat patients with whole brain irradiation. However, in September 1987 brain treatments were restarted and the trimmer bars computer file was used with the data corresponding to the prior source. As a consequence of the error, 33 patients undergoing radiation therapy to the brain received radiation doses that exceeded the prescribed dose by 75%. By the time the NRC was notified of the case, 20 patients had died either during treatment or after its conclusion [55].

Exercise to identify lessons and preventive actions from this event:

– What procedures does my radiotherapy department have for keeping files and data which are no longer in use inaccessible for treatment planning?
– Are these procedures understood and followed?
– What provisions are there to instruct new staff?

V.4. INCORRECT REPAIR FOLLOWED BY LACK OF COMMUNICATION (SPAIN, 1990)

Following instability of the radiation beam of a linear accelerator, the interlock system terminated the treatment of a patient. The problem had arisen at the beginning of a long holiday weekend. An engineer from the maintenance company who had been adjusting an adjacent ^{60}Co therapy unit was requested to 'have a look' at the accelerator. Over the long weekend the maintenance engineer made several unsuccessful attempts to identify the cause of the fault. Eventually he manipulated the energy control system to be able to restore the radiation beam.

Patient treatments resumed the following Monday. There were established procedures for notifying the hospital's maintenance engineer and medical physicists, but these were not followed. If they had been, beam tests would have been performed as required after any repair that could affect the dosimetry of the accelerator.

A control panel meter permanently indicated 36 MeV regardless of the energy selected; e.g. for an electron energy selection of 10 MeV the selector light button indicated 10 MeV but the control panel meter indicated 36 MeV. The radiotherapy technologists operating the accelerator pointed out the discrepancy between these two indications to the maintenance engineer, but it was assumed that the meter on the control panel was mechanically stuck on 36 MeV. The engineer did not realize that his attempts to restore the radiation beam had forced the beam energy control to its maximum.

The consequences of the mistake were relevant only to treatments with electron beams, for which the delivered dose and beam energy (and therefore the electron dose distribution) differed substantially from the selection made at the accelerator's operating console. The penetration of the electron beams was that of 36 MeV electrons, even when smaller penetration depths were intended by selection of low energies; therefore, large doses were delivered to tissues deeper than intended. The dose delivered was also much higher than intended due to the 'focusing' effect of electrons into a smaller cross-section of the beam. The dose rate was between 3 and 7 times higher than intended, depending on the selected energy.

The subsequent court proceedings [56] established that during the 10 day duration of the malfunction 27 patients had been treated with electrons; 15 died with radiation as the primary cause, two died with radiation as a major contributor to their deaths and others had major disabilities. The main radiation effects contributing to death were respiratory insufficiency, bronchopneumonia, radiation myelopathy, oesophageal stenosis and perforation, renal insufficiency, cervical hemorrhage and arteriosclerosis. Major effects on patients who were alive at the time of the court decision were paralysis and severe disability.

Exercise to identify lessons and preventive actions from this event:

For manufacturers and maintenance companies:

- What procedures do I have to train and document the training of maintenance staff?
- Does the document identify equipment components and systems which were and were not included in the training and the instructions to refrain from servicing equipment for which they have not been trained?
- What provisions do I have to entrust access to safety critical adjustments and components only to maintenance engineers instructed specifically on the safety aspects of these components?

- What provisions do I have to prevent irradiation in clinical mode when some interlocks are disabled?
- What instructions and warnings to maintenance staff on handling interlocks do I have?

For radiotherapy departments:

- What procedures does my radiotherapy department have for handing over the equipment to maintenance engineers and for returning it to service for treatments?
- Do these procedures include advising the medical physicist to check whether maintenance has influenced beam parameters before treatment?
- Which procedures communicate unusual displays on equipment and assign responsibility for completely clarifying them before treatments are continued or resumed?

V.5. LACK OF PROCEDURES FOR ACCEPTANCE OF A TPS (UK, 1982–1991)

Until 1982, manual calculations were the only method available to hospitals for treatment planning. Treatments with a linear accelerator were usually performed with a fixed source–skin distance (SSD) of 100 cm. For treatments requiring an SSD other than the usual 100 cm a procedure had been established based on correction factors for varying distances, which radiotherapy technologists applied manually.

A computerized TPS was acquired in the autumn of 1982. This allowed isocentric treatments on a more frequent basis. Since isocentric techniques are carried out with distances to the skin shorter than the usual 100 cm SSD, for the first isocentric treatment the technologists (with the acquiescence of the medical physicist) used the distance correction factors previously established. Every subsequent isocentric treatment applied this manual calculation for distance correction.

It had not been realized that the computer planning software already incorporated corrections for varying SSD. The correction was therefore duplicated when the radiotherapy technologists continued to manually apply distance correction for distances other than 100 cm. Since no quality control measurements were made, the error was not detected and the practice of manually applying the distance correction factors every time that an isocentric treatment was used remained in place. At the end of 1991, almost a decade later, the error was discovered.

The error delivered radiation doses lower than the prescriptions in all isocentric treatments during that period of time, which involved 1045 patients. A study by Ash and Bates [26] of these patients concluded that 492 patients had developed local recurrences, possibly as a result of the underdose. For 189 patients it was not possible to be certain about the effects of the underdose. For 320 patients, it was considered unlikely that they had suffered any effects from the underdose and in the remaining 44 patients the outcome was unknown.

An independent inquiry into the case found the following factors contributing to this accidental exposure: the responsibility and authority of the medical physicist was not clearly defined; the number of staff members was insufficient for the number of patients treated and, especially, the shortage of medical physicists led to overwork; a systematic QA programme that included commissioning following the introduction of new equipment was not in place; and the correction procedure that had been followed for nearly a decade had never been put down in writing.

Exercise to identify lessons and preventive actions from this event:

- What provisions does the QA programme have to include training staff in the proper understanding and use of new equipment, in particular TPS?
- What provisions does my radiotherapy department have for a systematic and comprehensive testing and commissioning treatment planning system?
- Is the responsibility to write procedures related to the use of equipment, and in particular TPS, as well as to follow the procedures, clearly defined?
- What procedure is there for manual checks of treatment time or manual monitoring of units?

V.6. MALFUNCTION OF BRACHYTHERAPY HDR EQUIPMENT (USA, 1992)

A patient was to be treated using an HDR brachytherapy unit equipped with a 16 GBq ^{192}Ir source. The prescribed dose was 18 Gy in three fractions. Five catheters were placed in the tumour and the source was to be stepped through the pre-programmed positions in each catheter. During the first fraction the radiation oncologist experienced difficulties in positioning the source into the fifth catheter and decided to retract the source. The source became detached from the driving mechanism while still inside the patient. The staff disregarded an alarm from an external area radiation monitor because the console of the brachytherapy unit indicated ‘safe’ and because there had been previous false alarms from the monitor. All three technologists and one physician who were attending the patient were aware of the alarm condition but none of them conducted a survey with the available portable radiation survey instrument [57].

The patient, with the source still in the catheter, was transported back to the nursing home. The source remained inside the patient for almost four days until the catheter containing the source fell out. The patient received a dose of 16 000 Gy at 1 cm distance from the source, instead of the prescribed 18 Gy. The nursing home staff disposed of the catheter in an area used to store non-radioactive medical waste and it was later removed by an incineration company. The source was discovered when it tripped a radiation monitor located at the incinerator.

The patient died shortly after the source was dislodged. The overexposure was the major contributing cause of death. The lost source also caused radiation exposure to 94 other individuals, including persons at the cancer clinic, nursing home, ambulance staff, and workers at the waste disposal company. A similar accidental exposure in another hospital was subsequently avoided because the medical physicist was aware of the first case and immediately recognized the problem. This emphasizes the importance of incident reporting and dissemination of the lessons learned.

Exercise to identify lessons and preventive actions from this event:

Manufacturers:

- What QA procedures do I have to test the source driving mechanism in stringent conditions, such as those in this case?
- Do these procedures include confirmation that the source has actually been returned to the safe and not just the driving mechanism?

Radiotherapy departments:

- What procedures does my radiotherapy department have for identifying responsibilities to completely clarify false alarms and contradictory displays?
- What procedures are there for monitoring the patient, clothes and the room with a portable monitor after removal of brachytherapy sources and before release of the patient?

V.7. BEAM MISCALIBRATION FOLLOWING THE EXCHANGE OF A ^{60}Co SOURCE (COSTA RICA, 1996)

A hospital had two ^{60}Co teletherapy units. Inconsistencies in dosimetry at the hospital had previously been detected by the IAEA/WHO TLD postal dose quality audit service. An external expert confirmed the deficiencies in dosimetry procedures and the lack of appropriate QA. These findings were reported to the hospital but were not acted upon. Insufficient education in medical physics, lack of an independent calibration, lack of a QA programme, the absence of documented procedures and lack of awareness at the hospital allowed mistakes to remain unnoticed.

After a source exchange in one of the ^{60}Co machines the medical physicist incorrectly calibrated the beam output, using a time of 30 s instead of 0.3 min (18 s) to determine the absorbed dose rate. This resulted in a 66% overestimation of the exposure time and therefore in an underestimation of the dose per unit time, which led to longer treatment times. The radiotherapy technologists questioned why patient treatment times remained about the same as with the old source when a new source was in place. The medical physicist told them that the delivered doses were correct according to his calculations. The radiation oncologist saw the patients only occasionally and ignored the high incidence of side effects such as vomiting, diarrhoea and rectal bleeding. Over a period of about one month, 114 patients received considerable overdoses, resulting in severe health effects including fatalities. Finally, another radiation oncologist at a different hospital who noted severe complications in his patients who had been treated on that machine detected the accidental exposure.

The dosimetry error was discovered and calculated by a physicist of the Pan American Health Organization (PAHO), assisted by local experts [58]. PAHO also sent medical consultants to evaluate patient reactions. Subsequently, an international medical team from the IAEA performed two assessments, one and two years after the accidental exposure [29]. Of the 51 patient deaths within two years of the accidental exposure, 13 were radiation related and 4 possibly radiation related. Of the 52 patients who where alive two years after the accidental exposure, 4 showed 'severe or catastrophic' effects due to radiation overexposure and 12 showed marked effects with a high risk of future effects. The radiation effects found in patients who survived two years after the accidental exposure were:

(a) Nervous system — brain: atrophy, necrosis, decreased cognitive function, headaches, mood alteration, seizures, decreased intellectual function; spinal cord: paralysis, quadriplegia and paraplegia.
(b) Skin: fibrosis, atrophy, contraction, induration, oedema, pigmentation, puritis, hypersensitivity, pain.
(c) Lower gastrointestinal tract: chronic or bloody diarrhoea, bowel stenosis, stricture, fibrosis, obstruction, fistula perforation.
(d) Bladder: dysuria, haematuria, contracture, incontinence.
(e) Vascular and lymphatic: stenosis, premature arteriosclerosis.

Exercise to identify lessons and preventive actions from this event:

– What local rules does my radiotherapy department have to avoid assigning duties to insufficiently trained staff?
– What provisions are there to stimulate working 'consciously' rather than mechanically (for example, awareness that a new source with higher

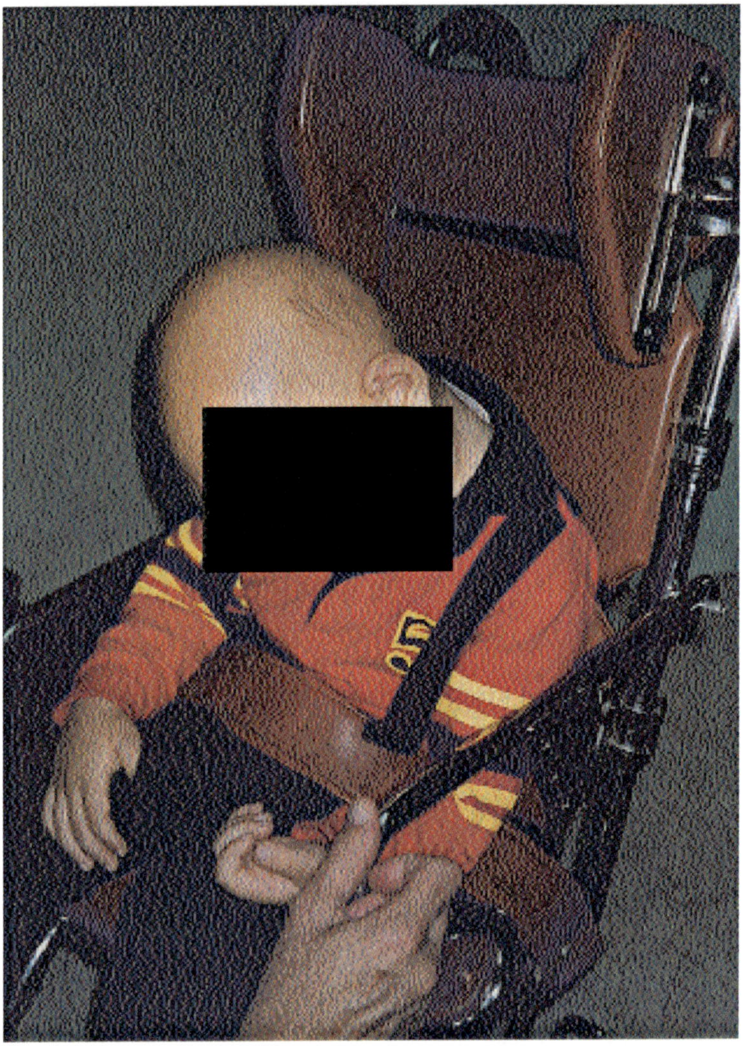

FIG. 5. A young child post-treatment and after radiotherapeutic overexposure for treatment of a brain tumour.

activity is expected to deliver a higher dose rate and require shorter treatment times)?

- What written procedures are there for calibration of beams and for an independent verification?
- Is participation in dose quality audits part of the QA programme?
- Would a mistake in the calibration of a beam be detected by independent checks before patients are incorrectly treated?

V.8. PROBLEMS WITH DATA ENTRY TO A TREATMENT PLANNING COMPUTER (PANAMA, 2000)

Shielding blocks are frequently used to protect normal tissue of patients undergoing radiotherapy, as was the case at the hospital in Panama. Data on the coordinates of shielding blocks are entered into the treatment planning computer, which calculates the dose distribution in patients and the treatment times.

Until August 2000, the practice had been to enter data for each shielding block separately. The TPS has a limitation on the number of shielding blocks for which data can be entered in this way. The practice was modified as of August 2000 to overcome this limitation for those treatments for which radiation oncologists prescribed pelvis fields with five shielding blocks, four in the corners and in some cases one more over the scars of hysterectomized patients treated for gynaecological cancers. The approach taken was to enter several single blocks as if they were one single block.[26] However, this approach caused the TPS to calculate incorrect radiation doses and, consequently, incorrect treatment times. Twenty-eight patients were affected by the incorrect doses, which were as much as twice the intended values.

An evaluation by the Pan American Health Organization (PAHO) and by a panel of US experts contracted by the hospital, and an IAEA/WHO investigation team [59], found that it was possible to enter data for several shielding blocks as a single block in different ways; and that for some ways of entering the data, which were accepted by the TPS, the output values were calculated incorrectly. However, whichever way was used, the computer produced a printout drawing that showed the treatment field and the shielding

---

[26] In this report, the expression 'entering data of shielding blocks as a single block' means that the blocks' coordinates were digitized by following the inner boundaries of the blocks, describing a loop, and then following the outer boundaries, describing another loop. At the end the transmission factor is entered once for all blocks.

blocks as if the data had been entered correctly. The isodose curves for a single treatment field were somewhat different, but for multiple treatment fields the differences were not so obvious. (It should be noted that for irradiation treatments in the pelvic region, which was the region of treatment for all the patients concerned, multiple treatment fields are always used at the Institute.[27]) These factors, together with an apparent omission of manual checking of computer calculations, resulted in the patients concerned being overexposed.

Of the 28 patients overexposed, eight had died by the time of the investigation in May 2001; and the team confirmed that five of these deaths were probably attributable to overexposure to radiation. Of the surviving 20 patients, most injuries were related to the bowel, with a number of patients suffering persistent bloody diarrhoea, necrosis (tissue death), ulceration and anaemia. The investigation concluded that about three quarters of the surviving 20 patients were expected to develop serious complications, which in some cases may ultimately prove fatal. In June 2002 a total of 19 patients had died.

Several characteristics of the TPS made the error more likely:

(a) It is questionable whether the information in the instructions is sufficiently clear to provide the user with detailed guidance as to how the blocks should be digitized;
(b) Several different ways of digitizing blocks were accepted by the computer;
(c) There was no warning on the computer screen when blocks were digitized in an unacceptable way, i.e. any way that was different from the one prescribed in the manual;
(d) When blocks were digitized incorrectly the TPS produced a diagram which was the same as that produced when data were entered correctly, thereby giving the impression that the calculated results were correct.

The modified data entry method was used without a verification test, i.e. a manual calculation of the treatment time for comparison with the computer calculated treatment time, or a simulation of a treatment by irradiating a water phantom and measuring the dose delivered. In spite of the treatment times being about twice those required for correct treatment, the error went

[27] 'Treatment field' is the term used in radiotherapy to denote the direction of the beam and the size and shape of its cross-section. Treatments in the pelvic region often require multiple treatment fields (that is, irradiation from different directions).

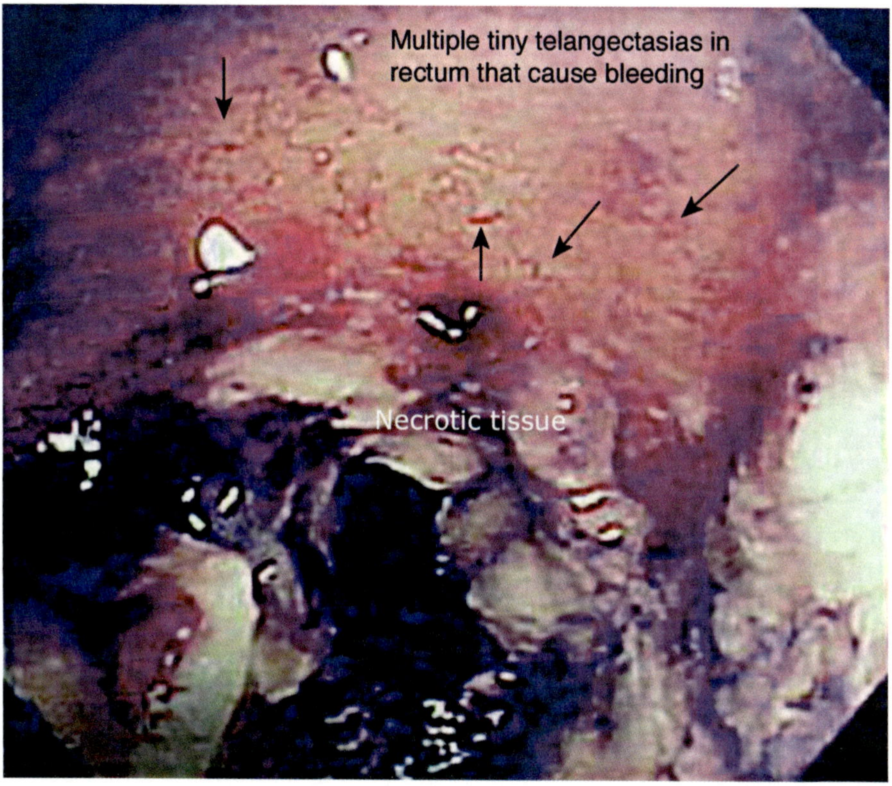

FIG. 6. Endoscopy of the colon of a patient showing necrosis and telangectasias.

unnoticed. Early symptoms of excessive exposure were noted in some of the irradiated patients but their seriousness was not understood, with the consequence that the accidental exposures went unnoticed for a number of months. The continued emergence of these symptoms, however, eventually led to detection of the accidental exposure. This was in March 2001.

Exercise to identify lessons and preventive actions from this event:

Manufacturers of TPS:

- Are instructions for users unambiguous with regard to how to enter data and operate the TPS and to refrain from other alternative ways of operating it?
- Are TPSs tested for possible incorrect operation?

- When there is an attempt to incorrectly operate the TPS, is there an interlock and a warning on screen and on the printed copy of the treatment plan?
- Do I offer sufficient advice to the users to avoid incorrect use because of lack of assistance?
- Is this advice readily available to the users in all countries where the TPS is purchased?

Radiotherapy departments:

- What kind of QA programme does my radiotherapy department have with provisions for testing and validating changes in procedures and for authorizing the modified procedures by the QA and radiation protection committees before they are applied?
- If treatment times in the treatment plan were substantially longer or shorter than usual, would my staff detect and report this anomaly?
- Would a follow-up be carried out until complete clarification is achieved before the patient is treated?

V.9. ACCIDENTAL EXPOSURE IN POLAND (2001)

On 27 February 2001 an accidental radiological overexposure occurred at the Bialystok Oncology Centre in Poland that affected five patients undergoing radiotherapy [60]. The accident resulted from a transitory loss of electric power, that caused an automatic shutdown of the Polish built NEPTUN 10P type linear accelerator. The power cut occurred during the radiation treatment of one of the affected patients. Following the restoration of electric power, the machine was restarted after its controls had been checked. The treatments were resumed and the initial patient and four other patients were treated.

Two patients experienced an itching and burning sensation during their irradiation. This caused the staff to halt further treatment. Subsequent dosimetry measurements revealed that the machine's output was significantly higher than expected. Further checks revealed that the dose monitoring system of the accelerator was not functioning properly and that one of the electronic components of the safety interlock system was damaged. Subsequently, all five patients developed local radiation injuries of varying severities.

An IAEA investigation team concluded that a fault affecting the beam monitoring system of the NEPTUN 10P accelerator had led to a large increase in the dose rate, even though the display indicated a lower value than normal. This was possible because a faulty diode had prevented the safety interlock from functioning. In addition, the limitation on the filament current for the electron gun was set at a high level so that the dose rate was many times higher than intended. The combination of these factors led to the substantially higher doses to patients.

The accidental exposure in Bialystok has demonstrated that two faults in two circuits may occur at the same time and lead to operation with an ineffective beam monitoring system. Moreover, the probability of a double fault was increased by the fact that an inoperative interlock (the diode) could go unnoticed until the second fault appeared in the beam monitoring system. In this situation the interlock could allow the start sequence to progress even if a second fault, a defective fuse in the power supply to the dose monitoring systems, rendered the system ineffective and led to a filament current being driven to its maximum.

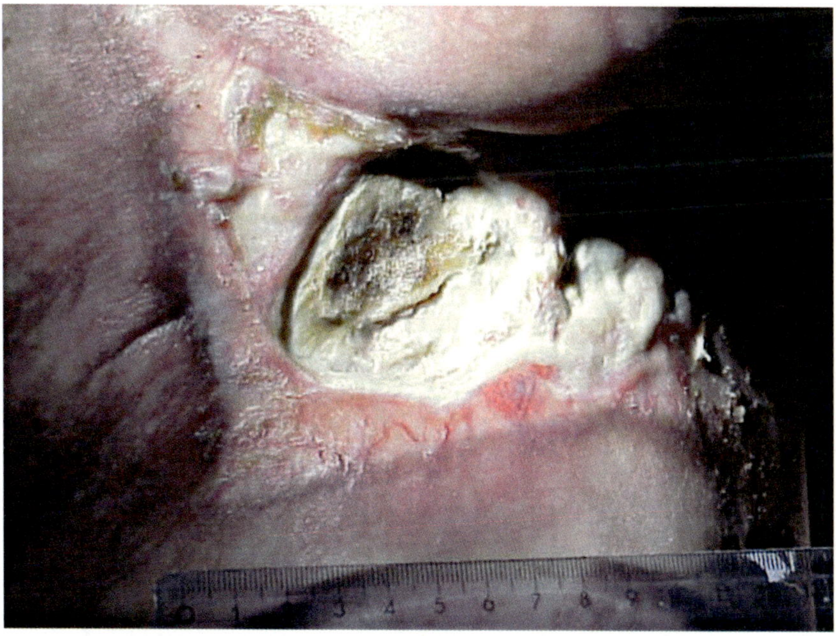

FIG. 7. Severe skin changes following a radiation burn. The wound covers a 14 cm × 8 cm area; in its centre lies an ulcer of full thickness covering a 5 cm × 4 cm area.

Exercise to identify lessons and preventive actions from this event:

Manufacturers and maintenance companies for radiotherapy equipment:

– Does the equipment conform to IEC or equivalent standards?[28]
– If the equipment was manufactured before current standards, has a review and assessment of safety been performed to reflect the new standards?[29]
– What procedures do I have to train and document the training of maintenance staff? Does the document identify equipment components and systems which were and were not included in the training, and instructions to refrain from servicing parts of equipment for which staff have not been trained?
– What provisions are there to entrust access to safety critical adjustments and components only to maintenance engineers instructed specifically on the safety associated with these components?
– What provisions are there to prevent irradiation in clinical mode when some interlocks are disabled?
– Which clear instructions and warnings to maintenance staff on the handling of interlocks exist?

Radiotherapy departments:

– What provisions does the QA programme include for relevant dosimetry checks after accelerator shutdowns due to power failures or any other unusual occurrence, such as an unusual indication in the dose rate display?
– Is the staff acquainted with these provisions?

---

[28] Current IEC standards [11] on medical electron accelerators require that "means to protect a possible overdose due to an absorbed dose rate more than twice the specified maximum, and limit the excess absorbed dose to less than 4 Gy...shall be tested between, or prior to, irradiations for their ability to function."

[29] Reference [60] states that "safety of existing equipment should be reviewed and reassessed for the need to increase safety to a level as close as practicable to that of the new standards. The improvements may be technological or procedural. The reassessment, however, should take into consideration all the implications of any change or modification. In relation with this event, issuing a warning notice with clear instructions requiring verification of relevant interlocks before each new irradiation and how to perform this verification could be necessary."

V.10. ADDITIONAL REMARKS

The above exercises on identifying lessons and preventive measures were only examples from major accidental exposures. For more comprehensive information the reader is encouraged to study Refs [27, 28], which contain a larger collection of information on a variety of cases. The problems identified, including the above major cases, can be summarized as follows:

Common to external beam and brachytherapy:

- Equipment not meeting IEC or equivalent national standards;
- Maintenance errors;
- Errors in the identification of patients and treatment sites;
- Conflicting signals and displays misinterpreted or not followed up;
- Communication errors, transmission of information and misunderstanding of prescriptions and protocols, or use of obsolete protocols;
- Use of obsolete files and forms which were still accessible.

External beam therapy:

- Errors in acceptance tests and commissioning or lack of tests of both radiation equipment and sources and TPSs;
- Errors in the calibration of radiotherapy beams;
- Errors in the preparation of tables and curves from which the treatment time is calculated;
- Errors in the use of TPSs for individual patients.

Brachytherapy:

- Using an incorrect source or incorrect units of source strength;
- Dislodging of HDR brachytherapy sources;
- Mistakes in source handling by nurses during brachytherapy treatment;
- Leakage of sealed sources;
- Sources left in patients and loss of radiation sources;

The following contributing factors allowed these errors to remain undetected until they became accidental medical exposures:

- Insufficient education of the radiation oncologist, medical physicist, radiotherapy technologist, maintenance engineers and brachytherapy nurses;

- Overloaded staff when new equipment was purchased or workload increased;
- Insufficient QA and lack of independent checks for safety critical activities, such as beam calibration;
- Lack of a programme for acceptance testing and commissioning;
- Lack of a maintenance programme;
- Poor, misunderstood or violated procedures;
- Lack of operating documents in a language understandable to the users;
- Inattention (environment prone to distraction);
- Inconsistent use of quantities and units.

In a number of the accidents there was a combination of several of the above contributing factors. Concurrent occurrence of several contributing factors may be indicative of a more general problem involving:

- Lack of commitment of the licensee (hospital administrators and managers of departments);
- Insufficiently educated or trained staff;
- Insufficient QA and defence in depth.

DEFINITIONS

The definitions given below may not necessarily conform to definitions adopted elsewhere for international usage.

absorbed dose. The fundamental dosimetric quantity D, defined as:

$$D = \frac{\mathrm{d}\varepsilon}{\mathrm{d}m}$$

where $\mathrm{d}\varepsilon$ is the mean energy imparted by ionizing radiation to matter in a volume element and $\mathrm{d}m$ is the mass of matter in the volume element. The energy can be averaged over any defined volume, the average dose being equal to the total energy imparted in the volume divided by the mass in the volume. The SI unit of absorbed dose is the joule per kilogram ($\mathrm{J \cdot kg^{-1}}$), termed the gray (Gy).

ambient dose equivalent, H*(*d*). The dose equivalent that would be produced by the corresponding aligned and expanded field in the ICRU sphere at a depth d on the radius opposing the direction of the aligned field. (It is used as a directly measurable proxy for effective dose for use in the monitoring of external exposure. A depth of $d = 10$ mm is recommended for strongly penetrating radiation.)

authorization. The granting by a regulatory body or other governmental body of written permission to perform specified activities.

dose constraint. A prospective and source related restriction on the individual dose delivered by the source which serves as a bound in the optimization of protection and safety of the source. For occupational exposures, the dose constraint is a source related value of individual dose used to limit the range of options considered in the process of optimization. For public exposure, the dose constraint is an upper bound on the annual doses that members of the public should receive from the planned operation of any controlled source. The exposure to which the dose constraint applies is the annual dose to any critical group, summed over all exposure pathways, arising from the predicted operation of the controlled source. The dose constraint for each source is intended to ensure that the sum of doses to the critical group from all controlled sources remains within the dose limit. There are no dose constraints for medical exposures of patients under treatment, except for persons exposed for medical research purposes or of persons, other than workers, who assist in the care, support or comfort of exposed patients.

effective dose. The quantity E, defined as a summation of the tissue equivalent doses, each multiplied by the appropriate tissue weighting factor:

$$E = \Sigma\, w_T\, H_T$$

where H_T is the equivalent dose in tissue $_T$ and W_T is the tissue weighting factor for tissue. From the definition of equivalent dose, it follows that:

$$E = \sum_T w_T \sum_R w_R D_{T,R}$$

where W_R is the radiation weighting factor for radiation R and $D_{T,R}$ the average absorbed dose in the organ or tissue T. The unit of effective dose is $J{\cdot}kg^{-1}$, termed the sievert (Sv).

emergency plan. A set of procedures to be implemented in the event of an accident.

employer. A legal person with recognized responsibility, commitment and duties towards a worker in his or her employment by virtue of a mutually agreed relationship. (A self-employed person is regarded as being both an employer and a worker.)

health professional. An individual who has been accredited through appropriate national procedures to practice a profession related to health (e.g. medicine, dentistry, chiropractics, paediatrics, nursing, medical physics, radiation and nuclear medical technology, radiopharmacy, occupational health).

health surveillance. Medical supervision intended to ensure the initial and continuous fitness of workers for their intended task.

intervention. Any action intended to reduce or avert exposure or the likelihood of exposure to sources which are not part of a controlled practice or which are out of control as a consequence of an accident.

kerma. The quantity K defined as:

$$K = \frac{dE_{tr}}{dm}$$

where dE_{tr} is the sum of the initial kinetic energies of all charged ionizing particles liberated by uncharged ionizing particles in a material of mass dm. The SI unit of kerma is the joule per kilogram ($J\cdot kg^{-1}$), termed the gray (Gy).

legal person. Any organization, corporation, partnership, firm, association, trust, estate, public or private institution, group, political or administrative entity or other persons designated in accordance with national legislation, who or which has responsibility and authority for any action having implications on protection or safety.

licence. A legal document issued by the regulatory body granting authorization to perform specified activities related to a facility or activity

licensee. The holder of a current licence.

medical exposure. Exposure incurred by patients as part of their own medical or dental diagnosis or treatment; by persons, other than those occupationally exposed, knowingly exposed while voluntarily helping in the support and comfort of patients; and by volunteers in a programme of biomedical research involving their exposure.

member of the public. In a general sense, any individual in the population except, for the purposes of the BSS, when subject to occupational or medical exposure. For the purpose of verifying compliance with the annual dose limit for public exposure, the representative individual in the relevant critical group.

normal exposure. An exposure which is expected to be received under normal operating conditions of a facility or activity, including possible minor mishaps that can be kept under control, i.e. during normal operation and anticipated operational occurrences.

notification. A document submitted to the regulatory body by a legal person to notify an intention to carry out a practice or other use of a source.

occupational exposure. All exposures of workers incurred in the course of their work, with the exception of exposures excluded from the BSS and exposures from practices or sources exempted by the BSS.

personal dose equivalent, Hp(*d*). The dose equivalent in soft tissue below a specified point on the body at the appropriate depth. (The relevant depths for the purposes of the BSS are generally $d = 10$ mm for strongly penetrating radiation and $d = 0.07$ mm for weakly penetrating radiation.)

potential exposure. Exposure that is not expected to be delivered with certainty but that may result from an accident involving a source or owing to an event or sequence of events of a probabilistic nature, including equipment failures and operating errors.

practice. Any human activity that introduces additional sources of exposure or additional exposure pathways, or extends exposure to additional people, or modifies the network of exposure pathways from existing sources, so as to increase the exposure or the likelihood of exposure of people or the number of people exposed.

protection and safety. The protection of people against exposure to ionizing radiation or radioactive materials and the safety of radiation sources, including the means for achieving this, and the means for preventing accidents and for mitigating the consequences of accidents should they occur.

public exposure. Exposure incurred by members of the public from radiation sources, excluding any occupational or medical exposure and the normal local natural background radiation, but including exposure from authorized sources and practices and from intervention situations.

qualified expert in radiotherapy physics. An individual who, by virtue of certification by appropriate boards or societies, professional licences or academic qualifications and experience, is duly recognized as having expertise in radiotherapy physics. Ideally such an individual should be a medical physicist with expertise in radiotherapy physics. The BSS require that for therapeutic uses of radiation (including teletherapy and brachytherapy), the calibration, dosimetry and QA requirements of the BSS be fulfilled by or under the supervision of a qualified expert in radiotherapy physics.

radiation generator. A device capable of generating ionizing radiation, such as X rays, neutrons, electrons or other charged particles, which may be used for scientific, industrial or medical purposes.

radiation protection officer. An individual technically competent in radiation protection matters relevant for a given type of practice who is designated by the registrant or licensee to oversee the application of the requirements of the BSS.

reference air kerma rate. The kerma rate of a source to air, in air, at a reference distance of one metre, corrected for air attenuation and scattering. This quantity is expressed in $mGy \cdot h^{-1}$ at 1 m.

regulatory body. An authority or system of authorities designated or otherwise recognized by a government for regulatory purposes in connection with protection and safety.

risk. A multi-attribute quantity expressing hazard, danger or chance of harmful or injurious consequences associated with actual or potential exposures. It relates to quantities such as the probability that specific deleterious consequences may arise, and the magnitude and character of such consequences.

safety assessment. A review of the aspects of design and operation of a source which are relevant to the protection of persons or the safety of the source, including the analysis of the provisions for safety and protection established in the design and operation of the source and the analysis of risks associated with normal conditions and accident situations.

safety culture. The assembly of characteristics and attitudes in organizations and individuals which establishes that, as an overriding priority, protection and safety issues receive the attention warranted by their significance.

sealed source. Radioactive material that is (a) permanently sealed in a capsule or (b) closely bounded and in a solid form. The capsule or material of a sealed source shall be strong enough to maintain leaktightness under the conditions of use and wear for which the source was designed, and also under foreseeable mishaps.

source. Anything that may cause radiation exposure, such as by emitting ionizing radiation or by releasing radioactive substances or materials. For example, materials emitting radon are sources in the environment, a sterilization gamma irradiation unit is a source for the practice of radiation preservation of food, an X ray unit may be a source for the practice of radiodiagnosis, and a nuclear power plant is a source for the practice of generating electricity by nuclear power. A complex or multiple installation situated at one location or site may, as appropriate, be considered a single source for the purposes of application of the BSS.

standards dosimetry laboratory. A laboratory designated by the relevant national authority for the purpose of developing, maintaining or improving primary or secondary standards for radiation dosimetry.

supplier. Any legal person to whom a registrant or licensee delegates duties, totally or partially, in relation to the design, manufacture, production or construction of a source. (An importer of a source is considered a supplier of the source.)

worker. Any person who works, whether full time, part time or temporarily, for an employer and who has recognized rights and duties in relation to occupational radiation protection. (A self-employed person is regarded as having the duties of both an employer and a worker.)

REFERENCES

[1] FOOD AND AGRICULTURE ORGANIZATION OF THE UNITED NATIONS, INTERNATIONAL ATOMIC ENERGY AGENCY, INTERNATIONAL LABOUR ORGANISATION, OECD NUCLEAR ENERGY AGENCY, PAN AMERICAN HEALTH ORGANIZATION, WORLD HEALTH ORGANIZATION, International Basic Safety Standards for Protection against Ionizing Radiation and for the Safety of Radiation Sources, Safety Series No. 115, IAEA, Vienna (1996).

[2] FOOD AND AGRICULTURE ORGANIZATION OF THE UNITED NATIONS, INTERNATIONAL ATOMIC ENERGY AGENCY, INTERNATIONAL LABOUR ORGANISATION, OECD NUCLEAR ENERGY AGENCY, PAN AMERICAN HEALTH ORGANIZATION, WORLD HEALTH ORGANIZATION, Radiation Protection and the Safety of Radiation Sources, Safety Series No. 120, IAEA, Vienna (1996).

[3] INTERNATIONAL ATOMIC ENERGY AGENCY, Legal and Governmental Infrastructure for Nuclear, Radiation, Radioactive Waste and Transport Safety, IAEA Safety Standards Series No. GS-R-1, IAEA, Vienna (2000).

[4] INTERNATIONAL ATOMIC ENERGY AGENCY, Organization and Implementation of a National Regulatory Infrastructure Governing Protection against Ionizing Radiation and the Safety of Radiation Sources, IAEA-TECDOC-1067, IAEA, Vienna (1999).

[5] COMISIÓN NACIONAL DE SEGURIDAD NUCLEAR Y SALVAGUARDIAS, Accidente por contaminación de con ^{60}Co, México 1984, CNSNS-IT-001, CNSNS, Mexico (1985).

[6] INTERNATIONAL ATOMIC ENERGY AGENCY, The Radiological Accident in Samut Prakarn, IAEA, Vienna (2002).

[7] INTERNATIONAL ATOMIC ENERGY AGENCY, Regulatory Control of Radiation Sources, IAEA Safety Standards Series No. GS-G-1.5, IAEA, Vienna (2004).

[8] INTERNATIONAL ATOMIC ENERGY AGENCY, Radiological Protection for Medical Exposure to Ionizing Radiation, IAEA Safety Standards Series No. RS-G-1.5, IAEA, Vienna (2002).

[9] WORLD HEALTH ORGANIZATION, Quality Assurance in Radiotherapy, WHO, Geneva (1988).

[10] INTERNATIONAL ATOMIC ENERGY AGENCY, Design and Implementation of a Radiotherapy Programme: Clinical, Medical Physics, Radiation Protection and Safety Aspects, IAEA-TECDOC-1040, IAEA, Vienna (1998).

[11] INTERNATIONAL ELECTROTECHNICAL COMMISSION, Particular Requirements for the Safety of Electron Accelerators in the Range of 1 to 50 MeV, Part 2, Publication IEC-60601-2-1, IEC, Geneva (1998).

[12] INTERNATIONAL ELECTROTECHNICAL COMMISSION, Particular Requirements for the Safety of Gamma Beam Therapy Equipment, Publication IEC-60601-2-11, IEC, Geneva (1997).

[13] INTERNATIONAL ELECTROTECHNICAL COMMISSION, Particular Requirements for the Safety of Remote-controlled Automatically-driven Gamma-ray Afterloading Equipment, Publication IEC-60601-2-17, IEC, Geneva (1998).

[14] INTERNATIONAL ELECTROTECHNICAL COMMISSION, Particular Safety of Therapeutic X-ray Generators, Part 2, Publication IEC-601-2-8, IEC, Geneva (1987).

[15] INTERNATIONAL ELECTROTECHNICAL COMMISSION, Particular Requirements for the Safety of Medical Radiotherapy Simulators, Part 2, Publication IEC-60601-2-29, IEC, Geneva (1999).

[16] INTERNATIONAL ELECTROTECHNICAL COMMISSION, Medical Electrical Equipment: Requirements for the Safety of Treatment Planning Systems, Publication IEC-62083, IEC, Geneva (2000).

[17] INTERNATIONAL ELECTROTECHNICAL COMMISSION, General Requirements for Safety — 4. Collateral Standard: Programmable Electrical Medical Systems, Publication IEC 60601-1-4, IEC, Geneva (1997).

[18] INTERNATIONAL ORGANIZATION FOR STANDARDIZATION, Basic Ionizing Radiation Symbol, ISO 361, ISO, Geneva (1975).

[19] INTERNATIONAL ATOMIC ENERGY AGENCY, Method for Developing Arrangements for Response to a Nuclear or Radiological Emergency: Updating IAEA-TECDOC-953, EPR-METHOD 2003, IAEA, Vienna (2003)

[20] INTERNATIONAL ORGANIZATION FOR STANDARDIZATION, Radiation Protection — Sealed Radioactive Sources — General Requirements and Classification, Rep. ISO 2919, ISO, Geneva (1998).

[21] INTERNATIONAL COMMISSION ON RADIOLOGICAL PROTECTION, Protection Against Ionizing Radiation from External Sources Used in Medicine, Publication 33, Pergamon Press, Oxford (1982).

[22] STEDEFORD, B., MORGAN, H.M., MAYLES, W.P.M., The Design of Radiotherapy Treatment Room Facilities, The Institute of Physics and Engineering in Medicine, York (1997).

[23] INTERNATIONAL ELECTROTECHNICAL COMMISSION, Guidelines for Radiotherapy Treatment Rooms, Publication IEC 61859, IEC, Geneva (1997).

[24] NATIONAL COUNCIL ON RADIATION PROTECTION AND MEASUREMENTS, Protection Against Radiation from Brachytherapy Sources, Rep. No. 40, NCRP, Bethesda, MD (1972).

[25] NATIONAL RADIATION PROTECTION BOARD, HEALTH AND SAFETY EXECUTIVE, DEPARTMENT OF HEALTH AND SOCIAL SERVICES (NORTHEN IRELAND), SCOTTISH HOME AND HEALTH DEPARTMENT, WELSH OFFICE, Guidance Notes for the Protection of Persons against Ionizing Radiation Arising from Medical and Dental Use, NRPB, Chilton (1988).

[26] ASH, D., BATES, T., Report on the clinical effects of inadvertent radiation underdosage in 1045 patients, Clinical Oncol. **6** (1994) 214–225.

[27] INTERNATIONAL ATOMIC ENERGY AGENCY, Lessons Learned from Accidental Exposures in Radiotherapy, Safety Reports Series No. 17, IAEA, Vienna (2000).
[28] INTERNATIONAL COMMISSION ON RADIOLOGICAL PROTECTION, Prevention of Accidental Exposure to Patients Undergoing Radiation Therapy, ICRP Publication 86, Pergamon Press, Oxford (2002).
[29] INTERNATIONAL ATOMIC ENERGY AGENCY, Accidental Overexposure of Radiotherapy Patients in San José, Costa Rica, IAEA, Vienna (1998).
[30] INTERNATIONAL ORGANIZATION FOR STANDARDIZATION, Radiation Protection — Sealed Sources — Leakage Test Methods, Rep. ISO 9978, ISO, Geneva (1992).
[31] SOCIEDAD ESPAÑOLA DE FÍSICA MÉDICA, The Accident of the Linear Accelerator in the "Hospital Clínico de Zaragoza", SEFM, Madrid (1991).
[32] INTERNATIONAL ATOMIC ENERGY AGENCY, Calibration of Photon and Beta Sources used in Brachytherapy, IAEA-TECDOC-1274, IAEA, Vienna (2002).
[33] INTERNATIONAL ATOMIC ENERGY AGENCY, Regulations for the Safe Transport of Radioactive Material, IAEA Safety Standards Series No. TS-R-1 (ST-1 Rev.), IAEA, Vienna (1996).
[34] INTERNATIONAL ATOMIC ENERGY AGENCY, Categorization of Radioactive Sources, IAEA-TECDOC-1344, IAEA, Vienna (2003).
[35] INTERNATIONAL ATOMIC ENERGY AGENCY, Occupational Radiation Protection, IAEA Safety Standards Series No. RS-G-1.1, IAEA, Vienna (1999).
[36] INTERNATIONAL ATOMIC ENERGY AGENCY, Assessment of Occupational Exposure due to External Sources of Radiation, IAEA Safety Standards Series No. RS-G-1.3, IAEA, Vienna (1999).
[37] INTERNATIONAL COMMISSION ON RADIOLOGICAL PROTECTION, Conversion Coefficients for Use in Radiation Protection against External Radiation, Publication 74, Pergamon Press, Oxford (1996).
[38] INTERNATIONAL COMMISSION ON RADIOLOGICAL PROTECTION, Protection and Safety in Medicine, Publication 73, Pergamon Press, Oxford (1996).
[39] INTERNATIONAL COMMISSION ON RADIOLOGICAL PROTECTION. Pregnancy and Medical Radiation, Publication 84, Pergamon Press, Oxford (2000).
[40] STOVAL, M., et al., Fetal dose from radiotherapy with photon beams: Report of the AAPM Radiation Therapy Task Group No. 36, Med. Phys. **22** (1995) 63–82.
[41] INTERNATIONAL ATOMIC ENERGY AGENCY, Absorbed Dose Determination in Photon and Electron Beams, 2nd edn, Technical Reports Series No. 277, IAEA, Vienna (1997).
[42] INTERNATIONAL ATOMIC ENERGY AGENCY, Absorbed Dose Determination in External Beam Radiotherapy, Technical Reports Series No. 398, IAEA, Vienna (2000).

[43] INTERNATIONAL ORGANIZATION FOR STANDARDIZATION, Radiation Protection — Sealed Radioactive Sources, General Requirements and Classification, Rep. ISO 2919, ISO, Geneva (1999).

[44] INTERNATIONAL ATOMIC ENERGY AGENCY, The Use of Plane Parallel Chambers in High Energy Electron and Photon Beams, Technical Reports Series No. 381, IAEA, Vienna (1997).

[45] INTERNATIONAL COMMISSION ON RADIATION UNITS AND MEASUREMENTS, Prescribing, Recording and Reporting Photon Beam Therapy, Rep. 50, ICRU, Bethesda, MD (1993).

[46] INTERNATIONAL COMMISSION ON RADIATION UNITS AND MEASUREMENTS, Prescribing, Recording and Reporting Photon Beam Therapy (Supplement to ICRU Rep. 50), Rep. 62, ICRU, Bethesda, MD (1999).

[47] AMERICAN ASSOCIATION OF PHYSICISTS IN MEDICINE, Comprehensive QA for radiation oncology, Report of the AAPM Radiation Therapy Committee Task Group 40, Med. Phys. **21** (1994) 1–38.

[48] INTERNATIONAL ATOMIC ENERGY AGENCY, Aspectos Físicos de la Garantía de Calidad en Radioterapia: Protocolo de Control de Calidad, IAEA-TECDOC-1151, IAEA, Vienna (2000).

[49] NUCLEAR REGULATORY COMMISSION, Consolidated Guidance about Materials Licensees, Rep. NUREG-1556, Vol. 9, Office of Standards Development, Washington, DC (1998).

[50] INTERNATIONAL ATOMIC ENERGY AGENCY, Preparedness and Response for a Nuclear or Radiological Emergency, Safety Standards Series No. GS-R-2, IAEA, Vienna (2002).

[51] INTERNATIONAL ATOMIC ENERGY AGENCY, Method for the Development of Emergency Response Preparedness for Nuclear or Radiological Accidents, IAEA-TECDOC-953, IAEA, Vienna (1997).

[52] COHEN, L., SCHULTHEISS, T.E., KENNAUGH, R.C., A radiation overdose incident: Initial data, Int. J. Radiat. Oncol. Biol. Phys. **33** (1995) 217–224.

[53] LEVESON, N., TURNER, C.S., An investigation of the Therac-25 accidents, IEEE Computer **26** 7 (1993) 18–19.

[54] RAWLINSON, J.A., "Report on the Therac-25", Proc. OCTRF/OCI Physicists Mtg Kingston, Canada (1987).

[55] NUCLEAR REGULATORY COMMISSION, Report to Congress on Abnormal Occurrences. AS-88-8. Multiple Medical Teletherapy Misadministration at Sacred Heart Hospital in Cumberland, Maryland, Rep. NUREG-0090, NRC, Washington, DC (1988, 1996).

[56] Account of the court proceedings and verdict for the case of the accelerator accident in Zaragoza (Spain), El Periódico, Madrid, 7 April 1993 (in Spanish).

[57] NUCLEAR REGULATORY COMMISSION, Report to Congress on Abnormal Occurences. 92-18. Loss of Iridium-192 Source and Medical Therapy Misadministration at Indiana Regional Cancer Center in Indiana, Pennsylvania, Rep. NUREG-0090, Vol. 15, No. 4, NRC, Washington, DC (1992).

[58] BORRÁS, C., "Radiation therapy overexposure", REMPAN 97 (Proc. 7th Coordination Mtg Rio de Janeiro, 1997), IRD, Paris (1997).
[59] INTERNATIONAL ATOMIC ENERGY AGENCY, Investigation of an Accidental Exposure of Radiotherapy Patients in Panama, IAEA, Vienna (2001).
[60] INTERNATIONAL ATOMIC ENERGY AGENCY, Accidental Overexposure of Radiotherapy Patients in Bialystok, IAEA, Vienna (2004).

CONTRIBUTORS TO DRAFTING AND REVIEW

Andreo, P.	International Atomic Energy Agency
Artus, F.	International Society for Radiation Oncology
Betancourt, A.	Centro Nacional de Seguridad Nuclear, Cuba
Bilbao, A.	International Atomic Energy Agency
Borrás, C.	International Organization for Medical Physics
Chomicki, O.	International Organization for Medical Physics
Cummings, B.	International Society for Radiation Oncology
Fuente, A.	Centro Nacional de Seguridad Nuclear, Cuba
Djeffal, S.	Ministère de l'Enseignement Supérieur et de la Recherche Scientifique, Algeria
Djermouni, B.	International Atomic Energy Agency
Domondon, D.	Philippine Nuclear Research Institute, Philippines
El Gueddari, B.K.	Ministère de la Santé Publique, Morocco
Gustafsson, M.	International Atomic Energy Agency
Ibbott, G.	University of Kentucky Medical Center, United States of America
Izewska, J.	International Atomic Energy Agency
Jiménez, P.	Pan American Health Organization
Kodia, E.	International Atomic Energy Agency
Larcher, A.M.	Autoridad Regulatoria Nuclear, Argentina
Leer, J.W.	International Society for Radiation Oncology
Levin, V.	International Atomic Energy Agency
Meghzifene, A.	International Atomic Energy Agency
Morkunas, G.	Radiation Protection Centre, Lithuania
Mrabit, K.	International Atomic Energy Agency

Niu, S.	International Labour Organisation, Geneva
Novotny, J.	Hospital Na Homolce, Czech Republic
Oliveira, S.M.	Comissão Nacional de Energia Nuclear, Brazil
Oliver, L.D.	Royal North Shore Hospital, Australia
Oresegun, M.	International Atomic Energy Agency
Ortiz López, P.	International Atomic Energy Agency
Ostensen, H.	World Health Organization
Pernicka, F.	International Atomic Energy Agency
Piccone, J.	Nuclear Regulatory Commission, United States of America
Pope, R.	International Atomic Energy Agency
Prinzio, R.	Comissão Nacional de Energia Nuclear, Brazil
Sabol, J.	International Atomic Energy Agency
Shortt, K.	International Atomic Energy Agency
Svensson, H.	International Society for Radiation Oncology
Wheatley, J.	International Atomic Energy Agency
Wrixon, A.D.	International Atomic Energy Agency

Consultants Meetings

Vienna, Austria: 10–14 August 1998, 22–26 February 1999, 22–26 November 1999, 13–15 December 1999

ARCAL–RCA–AFRA Meeting

Rio de Janeiro, Brazil: 21–25 June 1999

No. 20, August 2005

IAEA

International Atomic Energy Agency

Where to order IAEA publications

In the following countries IAEA publications may be purchased from the sources listed below, or from major local booksellers. Payment may be made in local currency or with UNESCO coupons.

Australia
DA Information Services, 648 Whitehorse Road, MITCHAM 3132
Telephone: +61 3 9210 7777 • Fax: +61 3 9210 7788
Email: service@dadirect.com.au • Web site: http://www.dadirect.com.au

Belgium
Jean de Lannoy, avenue du Roi 202, B-1190 Brussels
Telephone: +32 2 538 43 08 • Fax: +32 2 538 08 41
Email: jean.de.lannoy@infoboard.be • Web site: http://www.jean-de-lannoy.be

Canada
Bernan Associates, 4611-F Assembly Drive, Lanham, MD 20706-4391, USA
Telephone: 1-800-865-3457 • Fax: 1-800-865-3450
Email: order@bernan.com • Web site: http://www.bernan.com

Renouf Publishing Company Ltd., 1-5369 Canotek Rd., Ottawa, Ontario, K1J 9J3
Telephone: +613 745 2665 • Fax: +613 745 7660
Email: order.dept@renoufbooks.com • Web site: http://www.renoufbooks.com

China
IAEA Publications in Chinese: China Nuclear Energy Industry Corporation, Translation Section, P.O. Box 2103, Beijing

Czech Republic
Suweco CZ, S.R.O. Klecakova 347, 180 21 Praha 9
Telephone: +420 26603 5364 • Fax: +420 28482 1646
Email: nakup@suweco.cz • Web site: http://www.suweco.cz

Finland
Akateeminen Kirjakauppa, PL 128 (Keskuskatu 1), FIN-00101 Helsinki
Telephone: +358 9 121 41 • Fax: +358 9 121 4450
Email: akatilaus@akateeminen.com • Web site: http://www.akateeminen.com

France
Form-Edit, 5, rue Janssen, P.O. Box 25, F-75921 Paris Cedex 19
Telephone: +33 1 42 01 49 49 • Fax: +33 1 42 01 90 90 • Email: formedit@formedit.fr

Lavoisier SAS, 145 rue de Provigny, 94236 Cachan Cedex
Telephone: + 33 1 47 40 67 02 • Fax +33 1 47 40 67 02
Email: romuald.verrier@lavoisier.fr • Web site: http://www.lavoisier.fr

Germany
UNO-Verlag, Vertriebs- und Verlags GmbH, Am Hofgarten 10, D-53113 Bonn
Telephone: + 49 228 94 90 20 • Fax: +49 228 94 90 20 or +49 228 94 90 222
Email: bestellung@uno-verlag.de • Web site: http://www.uno-verlag.de

Hungary
Librotrade Ltd., Book Import, P.O. Box 126, H-1656 Budapest
Telephone: +36 1 257 7777 • Fax: +36 1 257 7472 • Email: books@librotrade.hu

India
Allied Publishers Group, 1st Floor, Dubash House, 15, J. N. Heredia Marg, Ballard Estate, Mumbai 400 001,
Telephone: +91 22 22617926/27 • Fax: +91 22 22617928
Email: alliedpl@vsnl.com • Web site: http://www.alliedpublishers.com

Bookwell, 2/72, Nirankari Colony, Delhi 110009
Telephone: +91 11 23268786, +91 11 23257264 • Fax: +91 11 23281315
Email: bookwell@vsnl.net

Italy
Libreria Scientifica Dott. Lucio di Biasio "AEIOU", Via Coronelli 6, I-20146 Milan
Telephone: +39 02 48 95 45 52 or 48 95 45 62 • Fax: +39 02 48 95 45 48

Japan
Maruzen Company, Ltd., 13-6 Nihonbashi, 3 chome, Chuo-ku, Tokyo 103-0027
Telephone: +81 3 3275 8582 • Fax: +81 3 3275 9072
Email: journal@maruzen.co.jp • Web site: http://www.maruzen.co.jp

Korea, Republic of
KINS Inc., Information Business Dept. Samho Bldg. 2nd Floor, 275-1 Yang Jae-dong SeoCho-G, Seoul 137-130
Telephone: +02 589 1740 • Fax: +02 589 1746
Email: sj8142@kins.co.kr • Web site: http://www.kins.co.kr

Netherlands
De Lindeboom Internationale Publicaties B.V., M.A. de Ruyterstraat 20A, NL-7482 BZ Haaksbergen
Telephone: +31 (0) 53 5740004 • Fax: +31 (0) 53 5729296
Email: books@delindeboom.com • Web site: http://www.delindeboom.com

Martinus Nijhoff International, Koraalrood 50, P.O. Box 1853, 2700 CZ Zoetermeer
Telephone: +31 793 684 400 • Fax: +31 793 615 698 • Email: info@nijhoff.nl • Web site: http://www.nijhoff.nl

Swets and Zeitlinger b.v., P.O. Box 830, 2160 SZ Lisse
Telephone: +31 252 435 111 • Fax: +31 252 415 888 • Email: infoho@swets.nl • Web site: http://www.swets.nl

New Zealand
DA Information Services, 648 Whitehorse Road, MITCHAM 3132, Australia
Telephone: +61 3 9210 7777 • Fax: +61 3 9210 7788
Email: service@dadirect.com.au • Web site: http://www.dadirect.com.au

Slovenia
Cankarjeva Zalozba d.d., Kopitarjeva 2, SI-1512 Ljubljana
Telephone: +386 1 432 31 44 • Fax: +386 1 230 14 35
Email: import.books@cankarjeva-z.si • Web site: http://www.cankarjeva-z.si/uvoz

Spain
Díaz de Santos, S.A., c/ Juan Bravo, 3A, E-28006 Madrid
Telephone: +34 91 781 94 80 • Fax: +34 91 575 55 63 • Email: compras@diazdesantos.es
carmela@diazdesantos.es • barcelona@diazdesantos.es • julio@diazdesantos.es
Web site: http://www.diazdesantos.es

United Kingdom
The Stationery Office Ltd, International Sales Agency, PO Box 29, Norwich, NR3 1 GN
Telephone (orders): +44 870 600 5552 • (enquiries): +44 207 873 8372 • Fax: +44 207 873 8203
Email (orders): book.orders@tso.co.uk • (enquiries): book.enquiries@tso.co.uk • Web site: http://www.tso.co.uk

On-line orders:
DELTA Int. Book Wholesalers Ltd., 39 Alexandra Road, Addlestone, Surrey, KT15 2PQ
Email: info@profbooks.com • Web site: http://www.profbooks.com

Books on the Environment:
Earthprint Ltd., P.O. Box 119, Stevenage SG1 4TP
Telephone: +44 1438748111 • Fax: +44 1438748844
Email: orders@earthprint.com • Web site: http://www.earthprint.com

United Nations (UN)
Dept. I004, Room DC2-0853, First Avenue at 46th Street, New York, N.Y. 10017, USA
Telephone: +800 253-9646 or +212 963-8302 • Fax: +212 963-3489
Email: publications@un.org • Web site: http://www.un.org

United States of America
Bernan Associates, 4611-F Assembly Drive, Lanham, MD 20706-4391
Telephone: 1-800-865-3457 • Fax: 1-800-865-3450
Email: order@bernan.com • Web site: http://www.bernan.com

Renouf Publishing Company Ltd., 812 Proctor Ave., Ogdensburg, NY, 13669
Telephone: +888 551 7470 (toll-free) • Fax: +888 568 8546 (toll-free)
Email: order.dept@renoufbooks.com • Web site: http://www.renoufbooks.com

Orders and requests for information may also be addressed directly to:

Sales and Promotion Unit, International Atomic Energy Agency
Wagramer Strasse 5, P.O. Box 100, A-1400 Vienna, Austria
Telephone: +43 1 2600 22529 (or 22530) • Fax: +43 1 2600 29302
Email: sales.publications@iaea.org • Web site: http://www.iaea.org/books

04-23831